**The Houghton Mifflin Series in Statistics
under the Editorship of Herman Chernoff**

LEO BREIMAN
Probability and Stochastic Processes: With a View Toward Applications
Statistics: With a View Toward Applications

Y. S. CHOW, HERBERT ROBBINS, AND DAVID SIEGMUND
Great Expectations: The Theory of Optimal Stopping

PAUL G. HOEL, SIDNEY C. PORT, AND CHARLES J. STONE
Introduction to Probability Theory
Introduction to Statistical Theory
Introduction to Stochastic Processes

PAUL F. LAZARSFELD AND NEIL W. HENRY
Latent Structure Analysis

GOTTFRIED E. NOETHER
Introduction to Statistics—A Fresh Approach

I. RICHARD SAVAGE
Statistics: Uncertainty and Behavior

Introduction to Statistical Theory

Paul G. Hoel
Sidney C. Port
Charles J. Stone
University of California, Los Angeles

HOUGHTON MIFFLIN COMPANY   **BOSTON**
Atlanta   Dallas   Geneva, Illinois   Hopewell, New Jersey
Palo Alto   London

LIBRARY OF CONGRESS CATALOG CARD NUMBER: 70-136172

ISBN: 0-395-04637-8

General Preface

This three-volume series grew out of a three-quarter course in probability, statistics, and stochastic processes taught for a number of years at UCLA. We felt a need for a series of books that would treat these subjects in a way that is well coordinated, but which would also give adequate emphasis to each subject as being interesting and useful on its own merits.

The first volume, *Introduction to Probability Theory*, presents the fundamental ideas of probability theory and also prepares the student both for courses in statistics and for further study in probability theory, including stochastic processes.

The second volume, *Introduction to Statistical Theory*, develops the basic theory of mathematical statistics in a systematic, unified manner. Together, the first two volumes contain the material that is often covered in a two-semester course in mathematical statistics.

The third volume, *Introduction to Stochastic Processes*, treats Markov chains, Poisson processes, birth and death processes, Gaussian processes, Brownian motion, and processes defined in terms of Brownian motion by means of elementary stochastic differential equations.

Preface

This book is designed for a one-semester course in mathematical statistics. It was written in close conjunction with *Introduction to Probability Theory*, the first volume of our three-volume series, and assumes that the student is acquainted with the material covered in a one-semester course in probability for which elementary calculus is a prerequisite.

The objective of this book is to present an elementary systematic treatment of mathematical statistics from a theoretical point of view. An attempt has been made to restrict consideration to important fundamental ideas and to describe these in some detail, so that the student will appreciate the motivation as well as the mathematics of the theory. Too often students who have finished a course in statistics come away with only a vague notion of the central ideas and methods of the subject. It is hoped that this text will uncover the unity and logical structure of statistical methods.

The theoretical development has been based on a few of the elementary notions of decision theory. This permits the treatment of Bayesian methods in addition to the more traditional methods; however, space did not permit the introduction of more than the most basic of these methods. The Bayesian techniques occur at the end of each chapter; therefore they can be omitted if time does not permit their inclusion.

One of the most important theorems and most useful techniques in statistics is concerned with testing the general linear hypothesis. This theorem is the foundation of numerous special tests and it can be applied to a host of important problems. A proof of the theorem is seldom presented at this elementary level; however, because of the importance of the theorem and because elementary calculus students are now receiving some training in matrix algebra, a proof based on simple algebraic and geometric techniques is presented. This material, which occurs in Chapter 5, is undoubtedly the most difficult part of the book, but its mastery is well worth the effort. For students who do not possess the necessary algebraic background, it is best to skip the proofs in this chapter and pass on to the applications.

Although elementary calculus suffices as a prerequisite for *Introduction to Probability Theory*, the present volume also assumes some elementary knowledge of matrix algebra. This knowledge will be needed in Chapter 4 as well as in Chapter 5, because the least squares theory in that chapter is presented by means of matrix notation and techniques. However it is only in Chapter 5 that a student needs to know anything more than the simplest notions of matrix algebra. A review of the matrix methods that are needed for these two chapters is presented in an appendix.

Some instructors may be surprised to discover that the concept of sufficiency is not introduced in this book. Sufficiency is very useful in developing statistical theory at an advanced level, but it would serve no useful purpose at this introductory stage. There are several other topics that are often found in introductory texts which are not included here. The justification for such omissions is that classroom time spent on such topics would leave insufficient time for an adequate discussion of the basic material.

The exercises at the end of each chapter are arranged according to the order in which the material of that chapter was presented. Problems of a computational nature occur first. Answers are given in an appendix.

Although this book was designed for a one-semester course meeting three times a week, it is sufficiently flexible in arrangement of material to adjust itself to a shorter course. This is easily accomplished either by omitting the chapter on nonparametric methods, or by omitting the material on Bayesian methods, or by omitting some of the proofs, or by a combination of such omissions. Sections that may be omitted are indicated by an asterisk. A somewhat longer course can be accommodated by including all the theory and spending more time on the exercises.

We would like to thank Frank Samaniego for obtaining answers to many of the exercises and Mrs. Gerry Formanack for her excellent typing.

Table of Contents

1 | *Basic Principles*

Many phenomena in the various sciences are governed by laws or relations of a stochastic nature. This implies that a probability model may be appropriate for representing the occurrence of the phenomenon. Such models were introduced and applied to games of chance and other physical experiments in Volume I, *Introduction to Probability Theory*. Although the physical sciences yield experiments that are likely to be more stable than those in the nonphysical sciences, and hence for which a probability model might seem to be more appropriate, such models can be just as appropriate in these other fields. The fundamental difference is that there may be more uncontrolled variables interfering with the variable being studied and leading to greater variability of it. The application of probability models to phenomena in these various sciences led to the development of methods that are now commonly called the methods of statistics.

Some simple typical problems that the methods of statistics are designed to solve are:

deciding on the basis of testing a few samples from a shipment of a certain drug whether the quality of the shipment is satisfactory,

predicting on the basis of a small poll what the voters' preferences are on a vital issue,

calculating on the basis of a high school student's record and the records of students who have gone to college what the chances are that he will be successful in college,

deciding on the basis of analyzing sonar signals whether a submarine is approaching.

Problems of this type can be formulated mathematically by considering the data that are to be used for making a decision as the observed values of a random variable X. The distribution of X is assumed to belong to a certain family of distributions, a particular member of which is specified when the value of a parameter θ is specified. The problem is to decide on the basis of the data which member, or members, of the family could represent the distribution of X. This is called *the problem of statistical inference*. For example, in the problem of determining voter preferences on an issue by means of a poll, the variable X may be treated as a discrete random variable with X assuming the value 1 or 0 corresponding to an individual's favoring or not favoring the issue, and with the parameter θ representing the unknown proportion of voters favoring the issue. A typical problem of statistical inference is to decide on the basis of a set of responses $(x_1, \ldots, x_n)$ obtained by a pollster whether the value of θ exceeds .60.

The methods of statistics are much broader in scope than the statistical inference problems illustrated here would suggest. They concern themselves also with such problems as how experiments should be conducted, what models are appropriate, and how information should be utilized. However, we shall be concerned almost exclusively with finding the best methods for making inferences about distributions of random variables. This means that we assume our random variable X possesses a given type of distribution depending upon an unknown parameter θ and that our objective is to draw some inference concerning θ. In most problems θ will be an explicit parameter of a probability density function $f(x \mid \theta)$; however, θ may be merely an index to distinguish different members of a family of such functions. The random variable X and the parameter θ may be vector variables with several components each; but in our discussion of basic principles they will be treated as one dimensional to simplify the exposition. The extension to vector variables will be considered in the next chapter. Typical probability models of this type are the binomial distribution with θ representing the probability of success in a single trial of an experiment, the Poisson distribution with θ representing the mean of the distribution, and the normal distribution with known variance and with θ representing the mean of the distribution. If both the mean and variance of a normal distribution were unknown, θ would represent the vector parameter (μ, σ).

In an inference problem such as the one where θ is the proportion of voters favoring an issue, the parameter θ is considered to be a fixed but unknown constant at the time the poll is taken. However, in some problems the parameter θ may be treated as a random variable with a known probability distribution. If so, the distribution will be assumed to be given by a density function, $\pi(\theta)$. The function $f(x \mid \theta)$ will then represent a conditional distribution with the variable θ fixed, and the joint distribution of X and θ will be given by the density $f(x, \theta) = \pi(\theta)f(x \mid \theta)$. Here the word density is used for both discrete and continuous random variables. This convention was used in Volume I and will be used throughout this book without further reminders. There is a slight inconsistency in notation here because a capital letter normally represents a random variable and the corresponding small letter its numerical value, whereas θ is being used here to represent both the random variable and its numerical value. When θ is treated as a random variable its probability distribution is called a *prior distribution*. This name arises because $\pi(\theta)$ is known prior to the experimentation that is carried out to make an inference concerning θ. An illustration of a problem for which θ might be treated as a random variable is that of deciding by means of testing a sample taken from a shipment whether the proportion θ of defectives in the shipment exceeds some tolerance proportion θ_0. Assume the purchasing firm receives such shipments regularly and has determined a distribution for θ from past shipment θ's. The θ for this shipment may be treated as the value, although unknown, of a random variable possessing this distribution. Most of the classical methods of statistical theory treat θ as a constant and rely only on a set of observed values of the random variable X for drawing inferences. This is partly because for many of the problems in the social and life sciences, from which much of statistical theory evolved, no $\pi(\theta)$ is available or appropriate. It would, of course, be a mistake to ignore prior information on θ if such information were available, even if it is not expressible in the form of a precise probability distribution.

To recapitulate, we shall assume that we are given a probability distribution of a random variable X that depends upon a parameter θ, and that we wish to make some inference concerning θ on the basis of some observed values of the random variable X and of a prior distribution for θ (if available).

1.1. Types of problems

Since an inference is to be made by means of a set of observed values $x_1, \ldots, x_n$ of the random variable X, it is necessary to introduce a function $d = d(x_1, \ldots, x_n)$ of those values for making the inference. Such a function is called a *decision function*. The nature of this function will depend upon the kind of inference concerning θ that is to be made. In the simplest problems we merely wish to know whether a certain proposition is true or false. For example, we might wish to know whether a shipment of drugs is up to quality specifications, whether a radar scanning has picked up a missile, or whether the number of children in a school district who suffer from malnutrition exceeds ten percent. We shall take a positive point of view in decision making by associating any decision with an action. Thus, in the preceding problems there will be two possible actions available, which will be denoted by a_1 and a_2, with a_1 corresponding to the decision of accepting the truth of the proposition and a_2 corresponding to its rejection. For a set of n observational or sample values, we will have an n-dimensional sample space. Since a decision function $d(x_1, \ldots, x_n)$ must determine for each point of this sample space whether action a_1 or a_2 is to be taken, such a function must separate the sample space into exactly two parts, one part consisting of those sample points for which a_1 will be taken and the other part consisting of points for which action a_2 will be taken. For example, if small values of X correspond to the truth of a proposition, a possible division of an n dimensional sample space might be to assign all points inside the sphere $x_1^2 + \cdots + x_n^2 = r^2$, where r is a suitably chosen constant, to a_1 and all other points to a_2. Problems of the preceding type in which there are only two possible actions are called *hypothesis testing problems*.

A more complicated decision making problem arises when there are more than two possible actions available. For example, suppose a certain region in Europe is known to have been inhabited by five different races of people. An archaeologist might wish to decide on the basis of bone measurements taken of a group of skeletons found in that region to which of the five races they belonged. Because of the simplicity of hypothesis testing procedures, problems involving more than two possible actions are sometimes incorrectly treated as hypothesis testing problems. For example, if a new drug is introduced as a cure for a disease, it is important to decide whether the drug is superior, inferior, or about equally effective in curing the disease; therefore it would be improper to treat it as a two action problem, in which one tests, for example, only whether the drug is superior or is equally effective in curing the disease. Problems in which there are a finite number, $k > 2$, possible actions available are called *multiple decision problems*. A decision function for such problems must divide the sample space into k parts, the ith part consisting of those sample points that are associated with taking action a_i, $i = 1, \ldots, k$.

A third class of problems arises when interest centers on trying to predict the value of the parameter θ and there are an infinite number of possibilities for θ. Thus if θ represents the proportion of voters who will vote for candidate A, it may be important to have a precise estimate of that proportion, rather than merely to decide whether it exceeds $1/2$. The decision function $d(x_1, \ldots, x_n)$ will then be a real-valued function whose range of values theoretically may be taken to be the interval $[0, 1]$. Problems of this type are called *estimation problems*.

As an illustration of how these three types of problems could arise in the same experimental situation, suppose that two new drugs for lowering blood pressure are to be compared for effectiveness. Let X denote the ratio of a patient's blood pressure after treatment to his blood pressure before treatment, and let θ represent the mean value of this random variable with respect to a class of patients. A typical problem of testing a hypothesis is to decide on the basis of experimentation with both drugs whether $\theta_1 \leq \theta_2$ or $\theta_1 > \theta_2$, where θ_1 and θ_2 correspond to the two drugs. From a practical point of view there is little point in preferring one drug to the other unless it shows a meaningful advantage. Thus, it might be more realistic to treat the problem as a multiple decision problem by considering the three possibilities $\theta_1 - \theta_2 \leq -\delta$, $-\delta < \theta_1 - \theta_2 < \delta$, $\theta_1 - \theta_2 \geq \delta$, where δ is the smallest difference that is considered practically useful. If the experiment yielded the decision, for whichever formulation was chosen, that the first drug is superior, then additional experimentation with this drug would be desirable so that an accurate estimate of θ_1 could be obtained.

1.2. The risk function

The success of a given decision function in accomplishing its objective needs to be measured in a numerical manner. If an experimenter is able to assign weights to the seriousness of making various incorrect decisions, these weights can serve to define a *loss function* $\mathscr{L}(\theta, a)$. This function is designed to numerically measure the penalty that arises from taking action a when θ is the true value of the parameter. For example, if we were given the observational values $x_1, \ldots, x_n$ of a normal variable with unknown mean θ, and we wished to use them to estimate θ, the action a would consist in stating that the value of θ is $d(x_1, \ldots, x_n)$. The magnitude of the error in this decision would be given by $|\theta - d(x_1, \ldots, x_n)|$. A typical loss function might then be $\mathscr{L}(\theta, a) = |\theta - a|$. To indicate the dependence of the loss function upon the decision function and the observational values, we will express it in the form $\mathscr{L}(\theta, d(x_1, \ldots, x_n))$. The name loss is attached to this function to indicate that the objective is to minimize $\mathscr{L}$. Since we wish to minimize our decision errors, it is clear that we would like $\mathscr{L}$ to be a function that decreases as the magnitude of the error decreases. The problem of how to choose $\mathscr{L}$ for the three types of problems discussed before will be considered in Section 1.4.

Problems arise in which the available decisions are qualitative in nature and for which it would be difficult to assign a numerical value to incorrect decisions. Thus, an individual might be faced with a choice of five color schemes for redecorating his house. If aesthetic considerations are as important as monetary ones in such a choice, it would be inconvenient to assign a loss function here. Problems of this type can be treated satisfactorily if one is able to assign a preference ordering to the possible choices by employing a numerical valued function, the utility function, that is based on this ordering. Even for problems of a quantitative nature, it is often necessary to introduce a utility function to express in quantitative form one's preferences among the various possibilities. We shall assume hereafter that if the problem is one in which it is necessary or desirable to introduce a utility function, then $\mathscr{L}$ represents that function.

Thus far the discussion has been on the basis of having available a set of observed values $x_1, \ldots, x_n$ of some random variable X and then trying to select a good function of those values. In measuring the effectiveness of any decision function, we must look at its overall performance and not just at how well it does for a single experiment. We therefore consider an experiment in which a set of n observations is to be taken of some random variable X. These potential observational values will be denoted by $X_1, \ldots, X_n$. If the observations are to be obtained by random sampling, then we know from the definition of random sampling in Volume I that these n random variables will be independently and identically distributed with the same distribution as that of X. A decision function $d = d(X_1, \ldots, X_n)$ is then a random variable, and therefore the loss $\mathscr{L} = \mathscr{L}(\theta, d(X_1, \ldots, X_n))$ is also a random variable. To measure the overall effectiveness of a decision function we calculate the expected value of the loss function and use it as our measure. This defines a new function called the risk function $\mathscr{R}$. Thus,

Definition 1 *The risk function $\mathscr{R}$ is given by the formula*

$$\mathscr{R}(\theta, d) = E_\theta \mathscr{L}(\theta, d(X_1, \ldots, X_n)),$$

where the expectation is taken with respect to the distribution of the random variables $X_1, \ldots, X_n$ with θ fixed.

Under random sampling and for the situation in which X possesses the density $f(x \mid \theta)$, the joint distribution of these n random variables is given by $\prod_{i=1}^n f(x_i \mid \theta)$. If X is a continuous random variable the risk function will then be given by

$$(1) \qquad \mathscr{R}(\theta, d) = \int \cdots \int \mathscr{L}(\theta, d(x_1, \ldots, x_n)) \prod_{i=1}^n f(x_i \mid \theta) \, dx_1 \cdots dx_n.$$

For a discrete random variable these integrals must be replaced by corresponding sums over all possible values of the x's. Since it is inconvenient to write out multiple integrals, an abbreviated notation will be used in which X will represent the basic random variable if a sample of size one is to be taken but will represent the vector random variable $X = (X_1, \ldots, X_n)$ if a sample of size $n > 1$ is to be taken. Then we may replace the multiple integration notation of (1) by the more compact representation

$$(2) \qquad \mathscr{R}(\theta, d) = \int \mathscr{L}(\theta, d(x)) f(x \mid \theta) \, dx$$

where now $f(x \mid \theta)$ denotes the density of the vector variable X and dx represents $dx_1 \cdots dx_n$.

Now suppose we wish to compare two decision functions d_1 and d_2 by means of their risk functions $\mathscr{R}(\theta, d_1)$ and $\mathscr{R}(\theta, d_2)$. This comparison is most easily made by means of their graphs. Consider the two sets of graphs shown in Figures 1 and 2 which represent two possible occurrences. It is clear in Figure 1 that decision function d_1 is better than d_2 because its risk function value is less than that of d_2 for each value of θ and our objective is to minimize the risk function. In Figure 2, however, neither function is superior to the other because for some values of θ the function d_1 is better than d_2 but for other values the advantage is reversed.

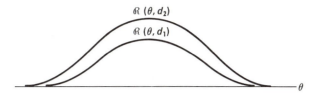

Figure 1

Unfortunately, the type of situation illustrated in Figure 1 rarely occurs, at least when the comparison is being made between decision functions that have been selected with intelligence. It would be difficult to make a choice between d_1 and d_2 for the more typical situation illustrated in Figure 2. In such cases, it is necessary to introduce some additional

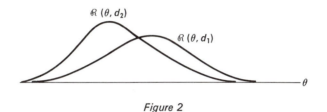

Figure 2

principle or criterion in order to arrive at a choice. One such principle that is quite popular is based on using the maximum value of the risk function as a criterion for comparison. If one risk function has a smaller maximum value than another, the decision function yielding the smaller maximum value is considered the better. If there is a decision function d whose risk function possesses a maximum value that is a minimum for all competing risk functions, d is called a best decision function in the minimax sense. Thus,

Definition 2   *The function d_0 is called a minimax decision function in the class D of decision functions if it satisfies*

$$\max_{\theta} \mathscr{R}(\theta, d_0) = \min_{d \in D} \max_{\theta} \mathscr{R}(\theta, d).$$

From Figure 2 it will be seen that d_1 is a better decision function than d_2 in the minimax sense, because its risk function clearly has a smaller maximum than the risk function for d_2. If the class D of decision functions consisted only of these two functions, then d_1 would be a minimax decision function relative to this class.

The advantage of introducing an additional principle, such as that of minimax, is that it reduces the comparison of decision functions to the comparison of real numbers. There are other principles that could be introduced here, but they will be discussed later when they are needed.

As an illustration of the preceding ideas, consider the problem of estimating the mean θ of a Poisson distribution on the basis of a single observed value X. Here $f(x \mid \theta) = (e^{-\theta}\theta^x)/x!$, $x = 0, 1, \ldots$. We shall choose the decision function $d(x) = cx$, where c is some positive constant, and we shall assume that the loss function is $\mathcal{L}(\theta, d) = (d - \theta)^2/\theta$. According to formula (1) the risk function is given by the sum

$$\mathcal{R}(\theta, d) = \sum_{x=0}^{\infty} \frac{(cx - \theta)^2}{\theta} \frac{e^{-\theta}\theta^x}{x!}.$$

The evaluation of $\mathcal{R}(\theta, d)$ is simpler, however, if the basic definition of $\mathcal{R}$ as the expected value of $\mathcal{L}$ is used and the properties of E that were derived in Volume I are employed; hence we calculate $E(cX - \theta)^2/\theta$. This expected value is most easily carried out by first writing $(cX - \theta)^2/\theta$ in the following form.

$$\frac{(cX - \theta)^2}{\theta} = \frac{c^2}{\theta}\left(X - \frac{\theta}{c}\right)^2 = \frac{c^2}{\theta}\left(X - \theta + \theta\left(1 - \frac{1}{c}\right)\right)^2$$

$$= \frac{c^2}{\theta}\left\{(X - \theta)^2 + 2\theta\left(1 - \frac{1}{c}\right)(X - \theta) + \theta^2\left(1 - \frac{1}{c}\right)^2\right\}.$$

Since X is a Poisson variable, we know from Volume I that the mean and variance of X are both equal to θ, and therefore that $E(X - \theta) = 0$ and $E(X - \theta)^2 = \theta$. Application of these facts to the preceding sum will yield the result

(3) $$\mathcal{R}(\theta, d) = \frac{c^2}{\theta}\left\{\theta + \theta^2\left(1 - \frac{1}{c}\right)^2\right\} = c^2 + \theta(c - 1)^2.$$

Now consider what value of c should be chosen to produce a good estimate of θ. If $c = 1$, the risk function has the constant value of 1. If $c \neq 1$, the risk function is a linear function of θ with a positive derivative, and therefore it will assume increasingly large values as θ becomes increasingly large. If there is no restriction on the value of θ, except that it must be positive, then $c = 1$ produces a minimax estimator in the class of estimators $d(X) = cX$. It is the only estimator in that class which yields a finite maximum for the risk function. The word estimator is used to denote a function of the random variables, whereas the word estimate denotes its numerical value.

1.3. Mean risk

If, in addition to the sample $X = (X_1, \ldots, X_n)$, a prior distribution for θ is available, then X and θ are both considered to be random variables.

The risk function $\mathcal{R}(\theta, d)$ is then a random variable because of its dependence on θ; therefore in order to measure the overall effectiveness of d as a decision function we calculate the expected value of the risk function and use that as our basis for making comparisons. This expected value, which is taken with respect to the prior distribution of θ, defines a new function called the mean risk, or Bayes risk, denoted by r. Thus,

Definition 3   *If θ is a continuous random variable having density $\pi(\theta)$, the mean risk is given by the formula*

(4) $$r(\pi, d) = E\,\mathcal{R}(\theta, d) = \int \mathcal{R}(\theta, d)\pi(\theta)\, d\theta.$$

If θ is a discrete variable, this integral must be replaced by a corresponding sum.

For a given decision function and prior distribution, $r(\pi, d)$ is a real number. Therefore decision functions can be compared by means of such numbers, just as in the case of minimax comparisons. If there exists a decision function d that minimizes $r(\pi, d)$ among all decision functions belonging to a class D of such functions, it should be called a best decision function in the mean risk sense; however, it is traditionally called the Bayes decision function with respect to the prior distribution π. Thus,

Definition 4   *A decision function d_0 is called a Bayes decision function with respect to the prior density π and the class D of decision functions if it satisfies*

$$r(\pi, d_0) = \min_{d \in D} r(\pi, d).$$

In the preceding definitions it has been assumed that we are dealing with functions that are integrable when integrals are involved and with functions that possess a maximum or minimum when the symbols max and min are involved. For students possessing sufficient knowledge to appreciate the difference, the symbols max and min should be replaced by sup and inf if more general definitions are desired.

As an illustration of how mean risks are calculated, consider the problem in the preceding section that was used as an illustration of estimating the mean of a Poisson distribution. Here we shall assume that the prior distribution of θ possesses the exponential density $\pi(\theta) = e^{-\theta}$, $\theta > 0$. Then, using the result obtained in (3), the mean risk becomes

$$r(\pi, d) = \int_0^\infty [c^2 + \theta(c - 1)^2]e^{-\theta}\, d\theta$$
$$= c^2 + (c - 1)^2 = 2c^2 - 2c + 1.$$

To obtain the Bayes decision function in the class of decision functions of the form $d(x) = cx$, it is necessary to choose the value of c that minimizes $r(\pi, d)$. Elementary calculus techniques will show that $r(\pi, d)$ is minimized by choosing $c = 1/2$; therefore $d(X) = X/2$ is the Bayes estimator corresponding to the prior density $\pi(\theta) = e^{-\theta}$.

The interesting feature of this result is that this Bayes estimate within the class of estimates of the form $d(x) = cx$, is only half as large as the minimax estimate. The prior density $\pi(\theta) = e^{-\theta}$ weights small values of θ more heavily than large values, and therefore might be expected to produce a smaller value than the estimator $d(X) = X$ which does not use this information.

1.4. Choice of loss functions

The choice of the loss function $\mathscr{L}(\theta, a)$ will depend upon the nature of the problem and upon the assessment of the seriousness of the various possible incorrect decisions. An experimenter may, for example, have rather definite views on whether large errors are exceedingly serious as compared to moderate size errors, and he can therefore assist in choosing an appropriate loss function. It seldom occurs, however, that the experimenter has any such strong views, and he will usually welcome any suggestions from the statistician. Fortunately, in many problems there can be considerable variation in the choice of a loss function without an appreciable change in the nature of the optimal decision function, particularly when the sample size is large.

If large errors were a very serious matter, we might choose the function $\mathscr{L}(\theta, a) = c(a - \theta)^2$, where c is a positive constant. If the loss were considered to be proportional to the size of the error, we would choose $\mathscr{L}(\theta, a) = c|a - \theta|$. A more general function that is capable of adaptation to various situations is the function $\mathscr{L}(\theta, a) = c(\theta)|a - \theta|^b$, where b is some positive constant and $c(\theta)$ is some positive function of θ (see Exercise 6).

Since we are concerned with developing a systematic theory and not with the practical details of applying such a theory, we shall restrict our exposition to the simplest loss functions that have found much favor in both statistical theory and practice. These seemingly rather specialized loss functions could have been generalized slightly without changing the theory, but such generalizations would complicate the notation.

For problems of testing a hypothesis we shall choose

$$\mathscr{L}(\theta, a) = \begin{cases} 0, & \text{if the correct decision is made} \\ 1, & \text{if the incorrect decision is made.} \end{cases}$$

For multiple decision problems we shall use the same loss function as for testing a hypothesis. That is, we assign $\mathscr{L}$ the value 0 for a correct decision and the value 1 for an incorrect decision, regardless of which of the $k - 1$ possible incorrect decisions is made.

For problems of estimation we shall choose

$$\mathscr{L}(\theta, a) = (a - \theta)^2.$$

This is known as quadratic loss or squared error loss. Since the action a to be taken in an estimation problem is deciding that the value of θ is the number $d(x_1, \ldots, x_n)$, the loss function here is normally written in the form

$$\mathscr{L}(\theta, d) = (d - \theta)^2,$$

where $d = d(x_1, \ldots, x_n)$ is the estimate of θ.

In the following chapters the theory of estimation will be studied first, followed by the theory of testing hypotheses. Very little material will be presented on multiple decision problems because the theory becomes rather involved. The simpler parts of that theory will be found in the Bayesian section of the hypothesis testing chapter.

Exercises

1 Given $f(x \mid \theta) = \dfrac{e^{-(1/2)(x-\theta)^2}}{\sqrt{2\pi}}$ and $\mathscr{L}(\theta, a) = (a - \theta)^2$, suppose a single observed value of X is to be taken. Calculate the value of $\mathscr{R}(\theta, d)$ if d is chosen to be the function $d(x) = cx$, where c is a constant.

2 For Exercise 1 determine whether there is a minimax decision function in the class of functions $d(x) = cx$.

3 Given $f(x \mid \theta) = \dbinom{2}{x} \theta^x (1 - \theta)^{2-x}$, $x = 0, 1, 2$, $0 < \theta < 1$, and $\mathscr{L}(\theta, a) = (a - \theta)^2$,
(a) calculate $\mathscr{R}(\theta, d)$ for $d(x) = x/2$;
(b) calculate $\mathscr{R}(\theta, d)$ for $d(x) = (x + 1)/4$.
(c) Which of these two functions is the minimax function with respect to the two functions?
(d) Is one decision function superior to the other?

4 Given the prior density $\pi(\theta) = 1/2$, $-1 < \theta < 1$,
(a) calculate the mean risk for Exercise 1;
(b) find the value of c which will produce a Bayes solution with respect to this prior distribution.

5 Given $f(x \mid \theta) = \binom{n}{x} \theta^x (1 - \theta)^{n-x}$, $x = 0, 1, \ldots, n$, $0 < \theta < 1$, and
$\mathscr{L}(\theta, a) = (a - \theta)^2$,
(a) calculate $\mathscr{R}(\theta, d)$ for $d(x) = x/n$.
(b) Given $\pi(\theta) = 1, 0 < \theta < 1$, calculate $r(\pi, d)$ for $d(x) = x/n$.

6 Work Exercise 5 if $\mathscr{L}(\theta, a) = (a - \theta)^2/\theta(1 - \theta)$.

7 Given $f(x \mid \theta) = \dfrac{e^{-\theta}\theta^x}{x!}$, $x = 0, 1, \ldots, \theta > 0$, and $\mathscr{L}(\theta, a) = (a - \theta)^2$,

(a) calculate $\mathscr{R}(\theta, d)$ for $d(x) = x$.

(b) Given the gamma density $\pi(\theta) = \dfrac{\lambda^\alpha \theta^{\alpha-1} e^{-\lambda\theta}}{\Gamma(\alpha)}$, $\alpha > 0$, $\lambda > 0$,

$\theta > 0$, calculate $r(\pi, d)$ for $d(x) = x$.

8 A coin is known to be biased with either $p = 1/4$ or $p = 3/4$, where p is
the probability of a head showing. A decision is to be made between
these two values on the basis of the outcome of two tosses of the coin.
Given $\mathscr{L}(\theta, a) = (a - p)^2$,
(a) calculate the value of $\mathscr{R}(\theta, d)$ for the following three decision
functions, where $d(X)$ must assume the value 1/4 or 3/4 for each
possible value of X and where X denotes the number of heads
obtained in the two tosses.

X	$d_1(X)$	$d_2(X)$	$d_3(X)$
0	1/4	3/4	3/4
1	1/4	3/4	1/4
2	1/4	3/4	1/4

(b) Which is the minimax decision function with respect to these three
functions?

9 Let $\theta = 0$ or 1, and let $f(x \mid 0) = 2^{-x}$, $x = 1, 2, \ldots$ and $f(x \mid 1) = 2^{-(x+1)}$, $x = 0, 1, \ldots$. Let $\mathscr{L}(0, 0) = \mathscr{L}(1, 1) = 0$ and $\mathscr{L}(1, 0) = \mathscr{L}(0, 1) = 1$. Consider the two decision functions

$$d_1(x) = \begin{cases} 1, & x = 0 \\ 0, & x \neq 0 \end{cases} \quad \text{and} \quad d_2(x) = \begin{cases} 0, & x \leq 1 \\ 1, & x > 1 \end{cases}.$$

Calculate $\mathscr{R}(\theta, d)$ for these two functions, and determine which one is
minimax with respect to the two functions.

2

Estimation

In this chapter we shall study methods for solving the third of the three basic problems of statistics that were discussed in the preceding chapter, namely that of estimation.

Since most of the problems that arise in statistical practice involve a random variable X that is either the discrete or the continuous type, we shall assume that X possesses a distribution given by a probability density function $f(x \mid \theta)$. As was stated in Chapter 1, the word density is used for both discrete and continuous random variables. We shall also assume that a sample $X_1, \ldots, X_n$ is to be taken of this random variable. The basic problem of estimation then is to choose a function $d = d(X_1, \ldots, X_n)$ that will best estimate the parameter θ of $f(x \mid \theta)$.

A function $d(X_1, \ldots, X_n)$ of the random variables $X_1, \ldots, X_n$ that is to be used to estimate θ is called an *estimator* of θ. The numerical value of this function that results when the observational values $x_1, \ldots, x_n$ of those random variables are inserted is called an *estimate* of θ; however, the word estimate is also commonly used for both the function d and its numerical value. We shall use the word estimator when it seems desirable to stress the fact that we are searching for a function to use for estimation purposes.

Since we are using the loss function $\mathscr{L}(\theta, d) = (d - \theta)^2$ for estimation problems and the risk function $\mathscr{R}(\theta, d)$ as the basis for comparing decision functions, it follows from

$$\mathscr{R}(\theta, d) = E(d(X_1, \ldots, X_n) - \theta)^2$$

that the basis for making comparisons will be the mean squared error.

It was pointed out in the preceding chapter that one is seldom able to find a function d in a class D of such functions that will minimize $\mathscr{R}$ for all values of θ. This is certainly true for estimation problems if the class D is not too restrictive. For example, suppose that we wish to estimate the mean θ of the normal density $f(x \mid \theta) = e^{-(1/2)(x-\theta)^2}/\sqrt{2\pi}$ by means of a random sample $X_1, \ldots, X_n$. A natural estimator to use here is the sample mean $\overline{X}$, in which case we would choose $d(X_1, \ldots, X_n) = (X_1 + \cdots + X_n)/n$ as our decision function. Another possible estimator is the number 10, which means that we would choose

13

$d(X_1, \ldots, X_n) = 10$ no matter what observational values are obtained. Now if θ happened to be equal to 10, this second estimator would yield a risk function value of zero, whereas $\overline{X}$ would yield the value $1/n$. The latter value arises from the fact learned in Volume I that if $EX = \theta$ and $E(X - \theta)^2 = \sigma^2$, then $E\overline{X} = \theta$ and $E(\overline{X} - \theta)^2 = \sigma^2/n$. Since $\sigma^2 = 1$ in our problem, it follows that $\mathscr{R}(\theta, \overline{X}) = 1/n$. Thus it is clear that we cannot expect to find an estimator that is best for all values of θ.

The minimax principle enables one to overcome the difficulty of comparing functions of θ for all values of θ by reducing the comparison to that of numbers, namely, the maximum value of the functions. We could define a best estimator to be a minimax estimator, and our basic problem would reduce to that of finding minimax estimates. There are, however, substantial objections to this approach. One such objection is that minimax estimates are usually difficult to find. Except for some of the simpler problems, the techniques required to find minimax estimates require ingenuity and sophisticated methods. Another objection is that the min-imax principle is a very conservative principle to employ, in that it will reject estimators that are excellent for most values of θ but which happen to have a larger maximum value than the minimax estimator. The minimax principle was designed for game theory problems, in which two equally intelligent players are competing and in which the expected loss for one of the players is given by the risk function. In order to fit statistical problems into such a game, the competing player is called nature and is supposed to have θ under his control, whereas the first player chooses the decision function d and, depending upon the choice of θ by nature, pays the amount $\mathscr{R}(\theta, d)$. The minimax principle tells the player to guard against the worst possible value of θ that nature can select, corresponding to any d that the player may choose, and then to select that d which minimizes the consequences of those worst values. In statistical problems for which θ is assumed to be some unknown constant there is no intelligent nature to manipulate θ so as to embarrass the statistician. Only the statistician is assumed to possess intelligence and use it to find a good estimator of θ.

For a simple illustration of the difference that occurs when using a minimax estimator rather than a familiar estimator, consider the problem of estimating the binomial distribution parameter p on the basis of a random sample of size n. The natural estimator is the success ratio $d = X/n$, where X denotes the total number of successes in the n trials of an experiment. Now we know from Volume I that $E(X/n) = p$; therefore $E(X/n - p)^2$ is the variance of the variable X/n. But we also know from Volume I that the variance of the variable X/n is $p(1 - p)/n$; hence, this is the value of $\mathscr{R}(p, d)$ for the estimator X/n. It can be shown by more advanced methods that the minimax estimator for this problem is given by the formula $d' \doteq (X + \sqrt{n}/2)/(n + \sqrt{n})$ and that its risk function has the value $(1/4)/(\sqrt{n} + 1)^2$. The graphs of these two risk functions are shown in Figure 1 for $n = 500$, with a convenient scale chosen on the vertical axis.

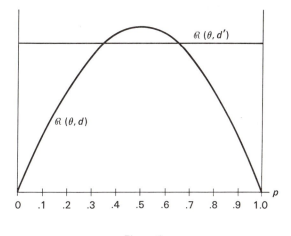

Figure 1

For values of p close to .5 the minimax estimator is superior; however, for values of $p < .35$ or $p > .65$ the success ratio estimator is superior. Since the difference in risk function values near .5 is not appreciable, whereas the difference for small or large values of p is considerable, the success ratio estimator would probably be preferred here. If there were reason to believe that the value of p is near .5, then, of course, the minimax estimator would be preferred. The minimax estimator shows up much better when the sample size is small, and it would then very likely be preferred to the success ratio estimator.

2.1. Unbiased estimates

In view of the difficulties involved in finding minimax estimates and because of their somewhat conservative virtues, we shall not use the minimax principle for defining a best estimator. Instead we shall restrict the class of estimators in the hopes that there will then be an estimator that will minimize the risk function for all values of θ. Such an estimator would be called a best estimator with respect to this restricted class. The restriction that will be placed on estimators is that of being *unbiased*. This property is defined as follows.

Definition 1 *An estimator $d = d(X_1, \ldots, X_n)$ is said to be an unbiased estimator of a parameter θ if $Ed(X_1, \ldots, X_n) = \theta$ for all values of θ.*

Since the mean squared error $E(d - \theta)^2$ becomes the variance of d when $E(d) = \theta$, the basis for comparing unbiased estimators is the variance of those estimators. This restriction is sufficient to insure that a best estimator will exist in many standard problems. Familiar examples of unbiased estimators are the success ratio X/n for the binomial parameter

p and the sample mean $\overline{X}$ for the normal parameter μ. The minimax estimator for the binomial parameter p, however, is easily shown to be biased.

2.2. Efficiency

If there exists an estimator that is unbiased and which possesses minimum variance among all unbiased estimators, it is called a best unbiased estimator. Such an estimator would enable us to judge any other unbiased estimator by comparing its variance with that of this best unbiased estimator. Such a comparison is most easily carried out by means of the ratio of the variances. If we let d^* represent the unbiased estimator that possesses minimum variance and let d denote an unbiased estimator that interests us, then we could employ the ratio $V(d^*)/V(d)$ as a measure of the efficiency of d as an estimator of θ. If an estimator d yielded an efficiency ratio of .80, it would be said to be 80 percent efficient. An estimator that is 100 percent efficient is usually called an *efficient estimator*. This ratio may, however, depend upon θ, in which case the efficiency would vary with θ.

Many unbiased estimators based on a random sample of size n possess variances of the form c/n, where c may depend on θ. If d and d^* were of this type and if d were, for example, 80 percent efficient, then the variances of d and d^* would be of the form c/n and $.8c/n$. The estimator d^* based on a sample of size 80 would be just as good as the estimator d based on a sample of size 100, because the ratio of their variances would be 1. Thus, for estimators of this type, the efficiency measure $V(d^*)/V(d)$ has the appealing property of determining the relative sample sizes needed to attain the same precision of estimation, as measured by the variance.

Although the preceding definition of the efficiency of an estimator is the natural one, it is seldom a workable definition because we have no method for finding an unbiased estimator that possesses minimum variance. Furthermore, there is no assurance that such an estimator even exists. If we do not know the value of the numerator in this ratio we can hardly discuss the efficiency of some other estimator. In view of the difficulty with this definition it is usually replaced by another closely related definition which can be applied to most problems.

Several mathematicians, working independently, were able under certain conditions to obtain a lower bound for the magnitude of the variance of any unbiased estimator. The density $f(x \mid \theta)$ of the random variable X must satisfy certain regularity conditions with respect to differentiation and integration, and the estimators being considered must possess a variance. If an unbiased estimator has a variance equal to this lower bound, we should be willing to call it a best unbiased estimator in our class of estimators. Any unbiased estimator whose variance attains this lower bound obviously has minimum variance, and therefore would be an efficient estimator in our class of estimators according to our first tentative proposal for measuring efficiency.

Since a lower bound of the variance is available, whereas the minimum possible variance $V(d^*)$ may not be, our measure of efficiency will be based on this lower bound. To show the dependence of this lower bound on θ it will be denoted by $B(\theta)$. As a substitute for $V(d^*)/V(d)$ we would then choose the ratio $B(\theta)/V(d)$.

The lower bound $B(\theta)$ that is needed for this definition and which was obtained by the mathematicians referred to earlier (Cramér, Frechet, Rao) is given by the formula

$$(1) \qquad B(\theta) = \frac{1}{nE\left(\dfrac{\partial \log f(X \mid \theta)}{\partial \theta}\right)^2} .$$

This formula is valid only for random samples, where n denotes the size of the sample. Since we shall be working exclusively with random samples, the regularity conditions are the only restrictions that need concern us. Our definition of efficiency now takes on the following operational form, wherein we restrict ourselves to estimation problems that satisfy the necessary regularity conditions and where δ represents any unbiased estimator possessing a second moment.

Definition 2 *The efficiency of an unbiased estimator δ is given by the formula*

$$(2) \qquad e(\delta) = \frac{1}{V(\delta)nE\left(\dfrac{\partial \log f(X \mid \theta)}{\partial \theta}\right)^2} .$$

Before deriving Formula (1) on which this definition is based, we shall apply the definition to a familiar problem for the purpose of illustrating its workability. Consider the problem of estimating the mean θ of a normal density with variance σ^2 by means of a random sample of size n. It is assumed that σ is known and therefore that it need not be estimated. Thus,

$$f(x \mid \theta) = \frac{\exp\{-(1/2)[(x - \theta)/\sigma]^2\}}{\sqrt{2\pi}\,\sigma} .$$

Hence,

$$\log f(x \mid \theta) = -\log \sqrt{2\pi}\,\sigma - \frac{1}{2\sigma^2}(x - \theta)^2$$

and

$$\frac{\partial \log f(x \mid \theta)}{\partial \theta} = \frac{x - \theta}{\sigma^2} .$$

Consequently,

$$E\left(\frac{\partial \log f(X \mid \theta)}{\partial \theta}\right)^2 = \frac{1}{\sigma^4} E(X - \theta)^2 = \frac{1}{\sigma^2}.$$

This last result follows from the fact that θ is the mean of X and therefore that $E(X - \theta)^2$ is the variance of X.

Now let us find the efficiency of the estimator $\overline{X}$ which we know is unbiased and whose variance is given by $V(\overline{X}) = \sigma^2/n$. Substitution of these two results into Formula (2) will yield an efficiency value of 1, showing that $\overline{X}$ is efficient. It is interesting to note that the efficiency of $\overline{X}$ is independent of the value of σ. Therefore, it is not necessary to know its value when claiming that $\overline{X}$ is efficient for estimating the mean of a normal density.

The lower bound (1) on which definition (2) is based can be obtained by elementary methods if the proper assumptions are made. This will be done by expressing (1) as a theorem and then proving the theorem for X a continuous random variable.

Theorem 1   *When certain regularity conditions are satisfied, the variance of any unbiased estimator δ of the parameter θ will satisfy the inequality*

$$V(\delta) \geq \frac{1}{nE\left(\dfrac{\partial \log f(X \mid \theta)}{\partial \theta}\right)^2}.$$

Proof.   Let $X_1, \ldots, X_n$ denote a random sample of X. The joint distribution of these random variables will then be given by the density $\prod_{i=1}^{n} f(x_i \mid \theta)$. If this density is denoted by L, and L is integrated over the entire sample space,

$$(3) \qquad \int \cdots \int L \, dx_1 \cdots dx_n = 1.$$

Since δ was assumed to be an unbiased estimator of θ, which means that $E\delta = \theta$, it follows that

$$(4) \qquad \int \cdots \int \delta L \, dx_1 \cdots dx_n = \theta.$$

The dependence of δ and L on $x_1, \ldots, x_n$ is not shown here in order to economize on writing.

Now we shall assume that the domain of definition of $f(x \mid \theta)$ does not depend upon θ in order that the limits of integration in the preceding integrals will not depend upon θ. Since L is a function of θ as well as of the x's, equations (3) and (4) are identities in θ, and therefore they may be differentiated with respect to θ to obtain additional identities. We shall assume that this differentiation may take place under the integral signs. Differentiation of identities (3) and (4) then yields the identities

$$\int \cdots \int \frac{\partial L}{\partial \theta} \, dx_1 \cdots dx_n = 0$$

and

$$\int \cdots \int \delta \frac{\partial L}{\partial \theta} \, dx_1 \cdots dx_n = 1.$$

For the purpose of recognizing these integrals, we shall let $t = \dfrac{\partial \log L}{\partial \theta}$ and use the relation $\dfrac{\partial L}{\partial \theta} = \dfrac{\partial \log L}{\partial \theta} \cdot L = tL$ to express these integrals in the form

(5) $$E(t) = \int \cdots \int tL \, dx_1 \cdots dx_n = 0$$

and

(6) $$E(\delta t) = \int \cdots \int \delta t L \, dx_1 \cdots dx_n = 1.$$

Since L is the density of the random variables $X_1, \ldots, X_n$ and δ and t are functions of those variables, these integrals represent the indicated expected values. Under the integral signs t and δ are understood to be functions of $x_1, \ldots, x_n$, but on the left they are understood to be the corresponding functions of the random variables $X_1, \ldots, X_n$.

Because δ and t are random variables, which we assume possess second moments, we may calculate their correlation coefficient ρ. We know from Volume I that

$$\rho = \frac{E(\delta t) - E(\delta) E(t)}{\sqrt{V(\delta) V(t)}}.$$

From results (5) and (6) this reduces to

$$\rho = \frac{1}{\sqrt{V(\delta) V(t)}}.$$

Since we know from Volume I that any correlation coefficient satisfies the inequality $\rho^2 \leq 1$, it follows from the preceding result that

(7) $$V(\delta) \geq \frac{1}{V(t)}.$$

But, because the sampling is random, the random variable

$$t = \sum_{i=1}^{n} \frac{\partial \log f(X_i \mid \theta)}{\partial \theta}$$

is the sum of n independent identically distributed random variables, each having the distribution of $\dfrac{\partial \log f(X \mid \theta)}{\partial \theta}$. It will be recalled from Volume I that the variance of the sum of independent variables is equal to the sum of the variances; therefore

$$(8) \qquad\qquad\qquad\qquad V(t) = n\sigma^2,$$

where σ^2 denotes the variance of $\dfrac{\partial \log f(X \mid \theta)}{\partial \theta}$. Furthermore, by the same type of reasoning with respect to expected values,

$$E(t) = nE \frac{\partial \log f(X \mid \theta)}{\partial \theta}.$$

But from (5) it follows that $E(t) = 0$ and therefore that

$$E \frac{\partial \log f(X \mid \theta)}{\partial \theta} = 0.$$

As a result, the variance of $\dfrac{\partial \log f(X \mid \theta)}{\partial \theta}$ is equal to its second moment and (8) reduces to

$$V(t) = nE \left(\frac{\partial \log f(X \mid \theta)}{\partial \theta} \right)^2.$$

When this result is inserted in (7) we obtain the desired lower bound of the theorem. ∎

If X is a discrete random variable, the preceding integrals must be replaced by corresponding sums. The regularity conditions referred to in the statement of this theorem are those that were assumed during the course of the proof. This theorem can be proved under much milder restrictions if more advanced methods are used, and therefore Definition 2 is not as restrictive as this proof might imply.

Throughout this book derivations and proofs will be carried out for continuous random variables only. The conclusions will be equally valid for discrete variables. It is merely necessary to replace integrals by sums to adapt such proofs to discrete variables.

As an illustration for a discrete variable, consider the problem of estimating the parameter θ for the Bernoulli random variable X whose density is given by $f(x \mid \theta) = \theta^x(1 - \theta)^{1-x}$, $x = 0, 1$, based on a random sample of size n. Here

$$\log f = x \log \theta + (1 - x) \log (1 - \theta),$$

$$\frac{\partial \log f}{\partial \theta} = \frac{x}{\theta} - \frac{1 - x}{1 - \theta} = \frac{x - \theta}{\theta(1 - \theta)},$$

and

$$E \left(\frac{\partial \log f}{\partial \theta} \right)^2 = \frac{1}{\theta^2(1 - \theta)^2} E(X - \theta)^2.$$

But $E(X - \theta)^2$ is the variance of X, which we know from earlier work is given by $\theta(1 - \theta)$; consequently

$$E\left(\frac{\partial \log f}{\partial \theta}\right)^2 = \frac{1}{\theta(1 - \theta)}.$$

The natural estimator to try here is the proportion of successes in n trials, namely $d = \sum_{i=1}^{n} X_i/n$. From Volume I we know that

$$E(d) = \theta \quad \text{and} \quad V(d) = \frac{\theta(1 - \theta)}{n}.$$

Thus, d is an unbiased estimator of θ possessing a second moment. Application of these results to (2) then shows that the sample proportion of successes in n trials is an efficient unbiased estimator of the probability of success in a single trial.

2.3. Asymptotic efficiency

In view of the fact that our only method for determining whether an efficient estimator exists is to find one that attains the lower bound in (2), we may not be able to tell how good an estimator is relative to other possible estimators if none attains this lower bound. Another serious disadvantage of our definition of efficiency is that it considers only unbiased estimators, and therefore excludes from consideration estimators that might have a smaller risk.

A definition of efficiency that is often employed when the preceding definition is considered inappropriate is based on the large sample properties of estimators. For the purpose of studying such properties, suppose we wish to estimate the quantity $\mu = EX$, where X possesses a non-normal density $f(x \mid \theta)$. Clearly μ will be some function of θ, and will be equal to θ only if θ is the mean of the distribution. Suppose further that we wish to compare the estimator $\bar{X}$ with some other estimator, call it Z, for relative efficiency as estimators of μ.

From the Central Limit Theorem of Volume I we know that under mild assumptions on $f(x \mid \theta)$ we can write

$$\lim_{n \to \infty} P\left(\frac{\sqrt{n}\,(\bar{X} - \mu)}{\sigma} < l\right) = \int_{-\infty}^{l} \frac{e^{-(t^2/2)}}{\sqrt{2\pi}}\, dt,$$

where σ^2 is the variance of X and l is any number. In general, μ and σ^2 are both functions of θ. When this type of limit exists, we say that $\bar{X}$ is asymptotically normally distributed with asymptotic mean μ and asymptotic variance σ^2/n. From a practical point of view, this means that $\bar{X}$ may be treated as a normal variable with mean μ and variance σ^2/n, provided that n is sufficiently large.

Now suppose that Z also possesses an asymptotic normal distribution and that it is of the form

$$\lim_{n \to \infty} P\left(\frac{\sqrt{n}\,(Z - \mu)}{c} < l\right) = \int_{-\infty}^{l} \frac{e^{-(t^2/2)}}{\sqrt{2\pi}}\, dt.$$

Here we would say that Z is asymptotically normally distributed with asymptotic mean μ and asymptotic variance c^2/n. As in the case of σ, the quantity c is some function of θ. For sufficiently large values of n we may treat Z as a normal variable with mean μ and variance c^2/n.

If two normal variables have the same mean μ but different variances, the probability of a sample value falling inside any interval of the type $(\mu - a, \mu + a)$ will be larger for the variable with the smaller variance. Since this will hold for every positive value of a, this means that the variable with the smaller variance will yield a more precise estimate of μ in this probability sense. Thus, for normal variables, the variance is an ideal measure of estimation efficiency. However, for variables that do not possess normal distributions, it does not follow that the variable with the smaller variance will yield a more precise estimate of μ.

Since $\overline{X}$ and Z may be treated as normal variables for large values of n, we may compare their relative asymptotic efficiency as estimators of μ by comparing their asymptotic variances σ^2/n and c^2/n. The estimator $\overline{X}$ would be said to be asymptotically more efficient than Z if $\sigma^2(\theta) \leq c^2(\theta)$ for all values of θ, with strict inequality holding for some value of θ.

For the purpose of obtaining a measure of asymptotic efficiency, rather than merely comparing two estimators, it suffices to modify Definition 2 by replacing $V(\delta)$ by the asymptotic variance of our estimator. Since δ was used to represent an unbiased estimator possessing a second moment and we are removing those restrictions here, we shall revert back to the symbol d. We now assume, however, that d possesses an asymptotic normal distribution with asymptotic mean μ. Thus,

Definition 3   *The asymptotic efficiency of an estimator d is given by*

(9)
$$e^*(d) = \frac{1}{c^2 E \left(\dfrac{\partial \log f(X\,|\,\theta)}{\partial \theta}\right)^2},$$

where c^2/n is the asymptotic variance of d.

It should be noted that the estimator d is not required to be unbiased, that the earlier regularity conditions are not imposed on $f(x\,|\,\theta)$, and that no loss function was introduced. Thus this definition has eliminated most of the possible objections to Definition 2; however, it has introduced its own disadvantages. Since this definition is concerned only with how estimators behave when n is large, it is of questionable value for small samples. The requirements that d possess an asymptotic normal distribution is a new restriction; however it is not as serious as it might appear because most of the natural estimators for standard problems do possess asymptotic normal distributions. If we restrict our estimators slightly more by requiring them to be what is called uniformly asymptotically normal, then under some suitable regularity conditions on $f(x \mid \theta)$ it can be shown that $e^*(d) \leq 1$; consequently any estimator for which $e^*(d) = 1$ is asymptotically efficient in this class of estimators.

In the case of estimating the mean of a normal variable it was found earlier that $\overline{X}$ is an efficient unbiased estimator according to Definition 2. Since $\overline{X}$ possesses a normal distribution for all values of n with mean μ and variance σ^2/n, it will obviously possess an asymptotic normal distribution with asymptotic mean μ and asymptotic variance σ^2/n. Furthermore, it will yield equality for (9) because it did so for (2); hence it is asymptotically efficient in the class of estimators discussed in the preceding paragraph. Since we are not requiring estimators to be unbiased here, this result shows that $\overline{X}$ is the best estimator in the asymptotic sense among biased as well as unbiased estimators for the mean of a normal variable.

In order to be able to apply Definition 3 to an estimator d it is necessary to derive the asymptotic variance of d. Fortunately, we shall not have to do this because certain methods of estimation have been shown to yield asymptotically efficient estimates. These will be considered in the next section.

2.4. Maximum likelihood estimation

One of the oldest known methods of estimation is based on choosing as an estimate of a parameter θ that value of θ which for a discrete variable maximizes the probability of obtaining the particular sample that was in fact obtained, and which for a continuous variable maximizes the probability density at the sample point. Given a density $f(x \mid \theta)$ and a set of random sample values $X_1 = x_1, \ldots, X_n = x_n$, we form the density function of the sample, denoted by $L(\theta)$,

$$L(\theta) = \prod_{i=1}^{n} f(x_i \mid \theta)$$

and call it the likelihood function. Since the x's are sample values of the random variable X, the quantity L is merely a function of the parameter θ. A maximum likelihood estimate is then a value of θ, which will be denoted by $\hat{\theta}$, that maximizes $L(\theta)$. In all the problems that will be encountered in this book there will be a unique value $\hat{\theta}$ that maximizes $L(\theta)$; consequently it will be assumed hereafter that there exists a unique $\hat{\theta}$. Since $\hat{\theta}$ will depend upon what particular sample values were obtained, we can express this dependence by writing $\hat{\theta} = \hat{\theta}(x_1, \ldots, x_n)$. If we wish to discuss a method of estimation that will be independent of what particular sample values are obtained, we must look upon $\hat{\theta}$ as the function $\hat{\theta}(X_1, \ldots, X_n)$ of the random variables $X_1, \ldots, X_n$, rather than of resulting numerical values. The method of maximum likelihood estimation is then defined as follows.

> **Definition 4**   *The maximum likelihood principle of estimation chooses the estimator $\hat{\theta} = \hat{\theta}(X_1, \ldots, X_n)$, where $\hat{\theta}$ is the function $\hat{\theta} = \hat{\theta}(x_1, \ldots, x_n)$ that maximizes the likelihood function $L(\theta) = \prod_{i=1}^n f(x_i \mid \theta)$ treated as a function of θ.*

Maximum likelihood estimates can often be obtained by the calculus techniques of maximizing a function that possesses a derivative. This method requires one to find the roots of $\dfrac{\partial L}{\partial \theta} = 0$ and then check to see whether one of those roots yields an absolute maximum of the function L. Some problems, however, cannot be solved by calculus methods, and others require refined numerical methods to solve the equation $\dfrac{\partial L}{\partial \theta} = 0$.

Maximum likelihood estimators possess some very desirable properties. Two of them are the following.

> **Property 1**   *If there exists an efficient unbiased estimator in the sense of Definition 2, then when certain regularity conditions are satisfied, the maximum likelihood method of estimation will produce it.*

This property is a very useful one because it gives us an operational method for searching for a best unbiased estimator. It is not necessary to try to guess a best estimator; it suffices to find the maximum likelihood estimator and then check to see whether it is unbiased and whether it produces equality in (2).

> **Property 2**   *Under certain regularity conditions, maximum likelihood estimators are asymptotically efficient in the sense of Definition 3.*

As in the case of Definition 2, the maximum likelihood method gives us an operational method for finding an asymptotically efficient estimator and therefore eliminates the necessity of finding asymptotic distributions.

Under the conditions of Property 2, it follows from that property that the maximum likelihood estimator $\hat{\theta}$ will possess an asymptotic normal distribution with mean θ and variance

$$c^2/n = 1/nE\left(\frac{\partial \log f(X \mid \theta)}{\partial \theta}\right)^2.$$

Thus, for large values of n we may treat $\hat{\theta}$ as a normal variable with mean θ and with a variance that can be calculated from a knowledge of $f(x \mid \theta)$.

As an illustration of the calculus method for finding maximum likelihood estimates, consider once more the problem of estimating the mean of a normal density. Here the likelihood function is

$$L(\theta) = \prod_{i=1}^{n} \frac{\exp\left[-\frac{1}{2}\left(\frac{x_i - \theta}{\sigma}\right)^2\right]}{\sqrt{2\pi}\,\sigma} = \frac{\exp\left[-\frac{1}{2}\sum_{i=1}^{n}\left(\frac{x_i - \theta}{\sigma}\right)^2\right]}{(2\pi)^{n/2}\sigma^n}.$$

Since it is easier to maximize the log $L(\theta)$ here than $L(\theta)$ and since these two functions will possess the same maximizing value of θ, we shall calculate $\dfrac{\partial \log L(\theta)}{\partial \theta}$. Thus

$$\log L(\theta) = -\log (2\pi)^{n/2}\sigma^n - \frac{1}{2}\sum_{i=1}^{n}\left(\frac{x_i - \theta}{\sigma}\right)^2,$$

and

$$\frac{\partial \log L(\theta)}{\partial \theta} = \frac{1}{\sigma^2}\sum_{i=1}^{n}(x_i - \theta).$$

Setting this derivative equal to zero and solving for θ, we conclude that $\hat{\theta} = \sum x_i/n = \bar{x}$. Since the second derivative here is negative, $\bar{x}$ produces a relative maximum value, which in turn is an absolute maximum value because $L(\theta)$ will approach 0 if θ approaches $+\infty$ or $-\infty$. Hence $\bar{X}$ is a maximum likelihood estimator of the mean of a normal density.

As an illustration for discrete variables, consider the problem of finding the maximum likelihood estimator for the parameter p of a Bernoulli distribution. Here $X = 1$ or 0, depending upon whether a success or failure occurred; hence if $\theta = p$ the density function is $f(x \mid \theta) = \theta^x(1 - \theta)^{1-x}$. Then

$$L(\theta) = \prod_{i=1}^{n} f(x_i \mid \theta) = \theta^{\sum x_i}(1 - \theta)^{n - \sum x_i}.$$

Consequently,

$$\log L(\theta) = \sum x_i \cdot \log \theta + \left(n - \sum x_i\right)\log (1 - \theta)$$

and

$$\frac{\partial \log L(\theta)}{\partial \theta} = \frac{\sum x_i}{\theta} - \frac{n - \sum x_i}{1 - \theta}.$$

Setting this derivative equal to zero and solving, we obtain the root $\hat{\theta} = \sum x_i/n$. Here also it is clear that $\sum x_i/n$ maximizes the likelihood function, either by the second derivative test or by looking at the graph of the likelihood function. Since $\sum x_i$ represents the total number of successes in the n trials of an experiment, it follows that the success ratio $\sum X_i/n$ is a maximum likelihood estimator of the probability of success in a single trial of the experiment.

The two definitions of efficiency that have been introduced in this chapter are probably the best known among such definitions. In recent years other definitions have been introduced that eliminate some of the restrictive assumptions made in our definitions, and hence which are of considerably wider applicability. Since they are more sophisticated mathematically than our definitions, it would be inadvisable to discuss them in this book.

2.5. Vector parameters

Thus far we have been concerned with finding best methods for estimating a single parameter. We encountered two parameters when estimating the mean of a normal density, but we did not try to estimate both of those parameters simultaneously. Since we usually wish to estimate all the unknown parameters that determine a distribution, we need to study the problem of how best to estimate a vector parameter that contains at least two components.

A generalization of the squared error loss function for estimating a single parameter would require us to introduce a quadratic form in the errors. For example, if two parameters θ_1 and θ_2 were being estimated by means of the estimators d_1 and d_2, our quadratic loss function would be of the form

$$\mathscr{L}(\theta, d) = (d_1 - \theta_1)^2 a + (d_1 - \theta_1)(d_2 - \theta_2)b + (d_2 - \theta_2)^2 c.$$

Unless higher dimensional quadratic loss functions are specialized considerably, the problem of finding best estimators becomes quite complicated. If there are k parameters to be estimated, the simplest such specialization is the choice

$$\mathscr{L}(\theta, d) = \sum_{i=1}^{k} (d_i - \theta_i)^2.$$

The risk function then becomes the sum of the mean squared errors, and if estimates are required to be unbiased the risk function reduces to the sum of the variances. In view of the difficulties that occur, we shall make no attempt to develop a systematic theory for this more general problem. Instead we shall be satisfied to apply the maximum likelihood technique to such problems in the hopes that it will produce good estimates for vector parameters just as it did for a single parameter. As a matter of fact it can be shown that maximum likelihood estimators do possess several optimal properties for vector parameter estimation. For example, among estimators that are asymptotically normally distributed, it can be shown that the maximum likelihood estimator is asymptotically efficient, where asymptotic efficiency for vector estimators is a natural generalization of its definition for a single parameter.

Suppose the random variable X possesses a density that depends upon k parameters $\theta_1, \ldots, \theta_k$. Let this density be denoted by $f(x \mid \theta_1, \ldots, \theta_k)$. Then the likelihood function corresponding to the random sample values $x_1, \ldots, x_n$ is defined by

$$L(\theta) = \prod_{i=1}^{n} f(x_i \mid \theta_1, \ldots, \theta_k).$$

The maximum likelihood estimate of the vector parameter θ is the set of values $\hat{\theta}_1, \ldots, \hat{\theta}_k$ that maximizes $L(\theta)$. Since these values will depend upon the particular x's obtained in the sample, they are functions of the x's. The corresponding maximum likelihood estimators will therefore be denoted by

$$\hat{\theta}_1 = \theta_1(X_1, \ldots, X_n)$$
$$\vdots$$
$$\hat{\theta}_k = \theta_k(X_1, \ldots, X_n).$$

As in the case of single parameter estimation, maximum likelihood estimates of vector parameters can often be obtained by calculus techniques, that is, by equating the partial derivatives of a function to zero. However, since it is considerably more difficult in higher dimensions to determine whether a calculus critical point is a maximizing point for the function, we shall often dispense with second derivative calculus tests and use other arguments to justify the claim that a unique critical point obtained from the likelihood equations is a maximizing point for the function.

As an illustration of the maximum likelihood method for vector parameter estimation, consider the problem of estimating the mean and variance of a normal density. For convenience of notation let the variance σ^2 be denoted by φ. Then the likelihood function based on a sample of size n is

$$L(\theta) = \frac{\exp\left[-\dfrac{1}{2\varphi} \displaystyle\sum_{i=1}^{n} (x_i - \mu)^2\right]}{(2\pi\varphi)^{n/2}}.$$

Taking logarithms and differentiating with respect to μ and φ gives

$$\log L = -\frac{n}{2} \log 2\pi - \frac{n}{2} \log \varphi - \frac{1}{2\varphi} \sum_{i=1}^{n} (x_i - \mu)^2,$$

$$\frac{\partial \log L}{\partial \mu} = \frac{1}{\varphi} \sum_{i=1}^{n} (x_i - \mu),$$

and

$$\frac{\partial \log L}{\partial \varphi} = -\frac{n}{2\varphi} + \frac{1}{2\varphi^2} \sum_{i=1}^{n} (x_i - \mu)^2.$$

Setting these derivatives equal to zero and solving the resulting equations, we obtain the estimates

(10) $\qquad\qquad \hat{\mu} = \bar{x} \quad$ and $\quad \hat{\varphi} = \dfrac{1}{n} \sum_{i=1}^{n} (x_i - \bar{x})^2.$

To show that these values do maximize log L, it suffices to show that they minimize $n \log \varphi + \dfrac{\sum (x_i - \mu)^2}{\varphi}$. But $\sum (x_i - \mu)^2$ will be minimized when $\mu = \bar{x}$, regardless of the value of $\varphi > 0$. It suffices therefore to choose φ to minimize $n \log \varphi + \dfrac{c^2}{\varphi}$, where $c^2 = \sum (x_i - \bar{x})^2$. The derivative of this function is $\dfrac{n}{\varphi} - \dfrac{c^2}{\varphi^2}$, which is negative for $\varphi < \dfrac{c^2}{n}$ and positive for $\varphi > \dfrac{c^2}{n}$; therefore $\varphi = \dfrac{c^2}{n}$ minimizes this function. The pair of values $\hat{\mu} = \bar{x}$ and $\hat{\varphi} = \sum (x_i - \bar{x})^2/n$ consequently maximize the likelihood function as claimed.

2.5.1. Sample moments. The estimates obtained in (10) are special cases of sample moments. For the purpose of explaining such moments let X be a discrete random variable that can assume only the values $x_1, \ldots, x_l$ with corresponding probabilities $f(x_1), \ldots, f(x_l)$. Then it will be recalled from Volume I that the kth moment of X was defined by

(11) $\qquad\qquad\qquad EX^k = \sum_{i=1}^{l} x_i^k f(x_i).$

The mean, or first moment, is usually denoted by μ. The kth moment of X about its mean is then defined by

(12) $\qquad\qquad\qquad E(X - \mu)^k = \sum_{i=1}^{l} (x_i - \mu)^k f(x_i).$

These definitions arose from considering a game in which a player wins the amount x_i^k if he scores x_i in the game. Let the game be played n times, and let $X_1, \ldots, X_n$ represent the outcomes. Then $\sum_{i=1}^{n} X_i^k/n$ will represent the average amount won. If the score x_i is obtained n_i times in these n games, this average can be expressed in the form

$$\frac{1}{n} \sum_{i=1}^{n} X_i^k = \sum_{i=1}^{l} x_i^k \frac{n_i}{n}.$$

Then as $n \to \infty$ we would expect n_i/n to approach the probability $f(x_i)$, and therefore we would expect the right side of this equality to approach the value given by the right side of (11). This motivation led to the definition expressed in (11).

In view of how the definition of the kth theoretical moment was obtained, it is natural to define *the kth sample moments* corresponding to (11) and (12) by the formulas

(13)
$$m'_k = \frac{1}{n} \sum_{i=1}^{n} X_i^k$$

and

(14)
$$m_k = \frac{1}{n} \sum_{i=1}^{n} (X_i - \bar{X})^k.$$

Thus, m'_k and m_k serve as sample estimates of EX^k and $E(X - \mu)^k$, based on a random sample of size n.

Although the preceding definitions arose from considering a discrete variable problem, they apply to continuous variables as well. Since every measuring device used to obtain samples is of limited accuracy, samples of a continuous random variable will behave like those of a discrete variable, and therefore there is no necessity for different definitions.

Returning now to the preceding section, a comparison of formulas (10) and (14) shows that the maximum likelihood estimates of the mean and variance of a normal density are given by the sample mean and the sample second moment about the sample mean, respectively.

In Section 2.2, in the problems of finding the efficiency of an unbiased estimator of a single parameter, it was assumed that any other parameters needed to specify $f(x \mid \theta)$ were known. For example, in estimating the mean of a normal density it was assumed that the variance was known. Similarly, in estimating the variance of a normal density it was assumed that the mean was known. In the case of estimating the mean, however, we were able to show that $\bar{X}$ is efficient whether σ^2 is known or not. Now we are in a position to consider the efficiency of an estimator even though some other parameters are unknown, because we can replace those other parameters by their maximum likelihood estimates. Thus, we may consider the problem of finding the efficiency of $\hat{\varphi}$ given in (10) as an estimate of the variance of a normal density when the mean is not known.

First, it is necessary to see whether $\hat{\varphi}$ is unbiased. This is accomplished by using algebraic manipulations to express $\hat{\varphi}$ in a form that will produce familiar expected values. The following expressions follow from such operations and the fact that $E(X_i - \mu)^2$ equals φ for random sampling.

$$E(\hat{\varphi}) = E \frac{1}{n} \sum (X_i - \bar{X})^2$$

$$= E \frac{1}{n} \sum [(X_i - \mu) - (\bar{X} -)\mu]^2$$

$$= E \frac{1}{n} \sum [(X_i - \mu)^2 - 2(X_i - \mu)(\bar{X} - \mu) + (\bar{X} - \mu)^2]$$

$$= E \frac{1}{n} \{ \sum (X_i - \mu)^2 - n(\overline{X} - \mu)^2 \}$$

$$= \frac{1}{n} \sum E(X_i - \mu)^2 - E(\overline{X} - \mu)^2$$

$$= \frac{1}{n} \sum \varphi - \frac{\varphi}{n}$$

$$= \frac{n-1}{n} \varphi.$$

Thus, it is seen that $\hat{\varphi}$ is biased; however, by using

$$d = \frac{n}{n-1} \hat{\varphi} = \frac{\sum (X_i - \overline{X})^2}{n-1}$$

as our estimator of φ, we will have an unbiased estimator whose efficiency can be determined by (2).

From Volume I we know that the square of a standard normal variable possesses a chi-square distribution with one degree of freedom, and that the sum of independent chi-square variables is a chi-square variable with degrees of freedom equal to the sum of the individual degrees of freedom. Therefore under random sampling from a normal variable, the expression $\frac{\sum_{i=1}^{n} (X_i - \mu)^2}{\sigma^2}$ possesses a chi-square distribution with n degrees of freedom. It was also shown that the mean and variance of a chi-square variable with n degrees of freedom are given by n and $2n$, respectively. In Chapter 5 it will be shown that $\frac{\sum_{i=1}^{n} (X_i - \overline{X})^2}{\sigma^2}$ also possess a chi-square distribution, but with $n - 1$ degrees of freedom. Using these facts it is not difficult to calculate the efficiency of the estimator $d = \frac{\sum (X_i - \overline{X})^2}{n-1}$ by means of (2) and show that it is $\frac{n-1}{n}$. When μ is known, the estimator $\frac{\sum (X_i - \mu)^2}{n}$, which is unbiased, possesses 100 percent efficiency; hence a slight loss in efficiency occurs when it is necessary to estimate μ. Both of these efficiency calculations will be omitted here because they are treated as exercises at the end of this chapter.

The sample second moment about the sample mean is usually called the sample variance; however, because the tradition of requiring estimates to be unbiased is still strong, the unbiased version is often called the sample variance. Since the standard deviation is the positive square root of the variance, the sample standard deviation is defined to be the square root of the sample variance. Here also, there is a choice of using either

$\sqrt{\sum (X_i - \overline{X})^2/n}$ or $\sqrt{\sum (X_i - \overline{X})^2/(n-1)}$ to define the sample standard deviation. Neither of these definitions, incidentally, produces an unbiased estimate of σ; it is only when estimating σ^2 that the criterion of unbiasedness suggests division by $n-1$ rather than n.

2.6.   Other methods

Problems arise for which the preceding methods are inappropriate because they give rise to computational difficulties or because the density $f(x \mid \theta)$ has not been specified. For example, we may wish to estimate the parameters α and β in the gamma density given by $f(x \mid \theta) = \dfrac{e^{-x/\beta} x^{\alpha-1}}{\beta^{\alpha} \Gamma(\alpha)}$. Application of the maximum likelihood method will produce two equations that are impossible to solve except by sophisticated numerical methods. One device that will overcome this difficulty is to calculate the mean and variance of the distribution, which will involve α and β, and equate these theoretical moments to the corresponding sample moments to produce two equations in α and β that may be solvable by elementary methods. Calculations here will show that $\mu = \alpha\beta$ and $\sigma^2 = \alpha\beta^2$; therefore the desired equations are

$$\alpha\beta = \overline{x} \quad \text{and} \quad \alpha\beta^2 = \frac{\sum (x_i - \overline{x})^2}{n}.$$

These yield the estimates

$$\alpha = \frac{n\overline{x}^2}{\sum (x_i - \overline{x})^2} \quad \text{and} \quad \beta = \frac{\sum (x_i - \overline{x})^2}{n\overline{x}}.$$

The technique employed in the preceding problem is called the *method of moments*. In general, if a density depends upon a vector parameter θ that possesses s components, the method of moments chooses as estimates of those parameters the values that satisfy the s equations

$$EX^k = m'_k, \quad k = 1, \ldots, s,$$

where m'_k is defined in (13).

Application of the method of moments to the problem of estimating the parameters μ and σ^2 for a normal density will yield the maximum likelihood estimates given in (10), because equating empirical and theoretical means and second moments about the origin is equivalent to equating empirical and theoretical means and second moments about the mean.

Not only does the method of moments often enable us to estimate parameters that are difficult to estimate by means of maximum likelihood techniques, but it is highly useful for estimating properties of a distribution for which no explicit density function is available. We may wish to know the values of the mean and variance of a random variable but have no knowledge as to the nature of its distribution; therefore we cannot discuss the problem of estimating the parameters of a density $f(x \mid \theta)$. If we merely assume that the distribution possesses two moments, we may use the first two sample moments to estimate the first two theoretical moments and thereby obtain an estimate of the mean and variance of the random variable being sampled. This type of problem is called *non-parametric* because we are not given an explicit density $f(x \mid \theta)$ depending on some parameter θ.

Although the method of moments is highly useful for problems that present difficulties of the preceding kind, there is no assurance that estimators obtained in this way possess high efficiency according to either of our definitions. If $f(x \mid \theta)$ is available, it is often possible to calculate the variance or asymptotic variance of such an estimator and then determine its efficiency.

This nonparametric method of estimation can be extended to more general functions than moments. Thus, suppose that X is a discrete variable which can assume only the values $x_1, \ldots, x_l$ with probabilities $f(x_1), \ldots, f(x_l)$, and that we wish to estimate $Eg(X)$ where g is some function that interests us. From Volume I we know that

$$(15) \qquad Eg(X) = \sum_{i=1}^{l} g(x_i) f(x_i).$$

This formula could be motivated in the same manner as the formula for the kth theoretical moments. Thus, suppose a game is to be played n times, and a player wins the amount $g(x_i)$ if he scores x_i in the game. Let $X_1, \ldots, X_n$ represent the outcomes; then $\dfrac{\sum_{i=1}^{n} g(X_i)}{n}$ will represent the average amount won. If the score x_l is obtained n_i times in these n games, this average can be expressed in the form

$$(16) \qquad \frac{1}{n} \sum_{i=1}^{n} g(X_i) = \sum_{i=1}^{l} g(x_i) \frac{n_i}{n}.$$

Now as $n \to \infty$ we would expect n_i/n to approach $f(x_i)$ and therefore we would expect the right side of (16) to approach the right side of (15). This leads to

$$(17) \qquad d(X) = \frac{1}{n} \sum_{i=1}^{n} g(X_i)$$

as our choice of estimator for $Eg(X)$.

Just as in the case of moments, this estimator suffices for both continuous and discrete variables. If X is a continuous random variable (15) must be replaced by

(18) $$Eg(X) = \int g(x)f(x)\,dx,$$

where $f(x)$ represents the density of X.

The estimate given by (17) is a nonparametric estimate for (15) or (18) because we are not given a density $f(x \mid \theta)$ depending upon some parameter θ. As a matter of fact, we are assuming no knowledge about the density $f(x)$ except that our functions g and f must be such that the integral in (18) exists. It is to be noted that the function $Eg(X)$ includes moments as a special case when the choice $g(X) = X^k$ is made.

The preceding methods can be generalized to include estimates of parameters and nonparametric functions of several random variables. For example, let X and Y be two correlated random variables with means and variances μ_x, μ_y, σ_x^2, and σ_y^2, and with correlation coefficient ρ; and suppose we wish to estimate ρ on the basis of the random sample values $(x_1, y_1), \ldots, (x_n, y_n)$. From Volume I we know that

$$\rho = \frac{E(X - \mu_x)(Y - \mu_y)}{\sqrt{V(X)V(Y)}}.$$

We may estimate the variances $V(X)$ and $V(Y)$ by means of

$$\sum_{i=1}^{n} \frac{(x_i - \bar{x})^2}{n} \quad \text{and} \quad \sum_{i=1}^{n} \frac{(y_i - \bar{y})^2}{n}.$$

As an estimate of the covariance $E(X - \mu_x)(Y - \mu_y)$ we shall choose

$$\sum_{i=1}^{n} \frac{(x_i - \bar{x})(y_i - \bar{y})}{n}.$$

As a result, our estimate of ρ, which we shall denote by r, assumes the form

$$r = \frac{\sum (x_i - \bar{x})(y_i - \bar{y})}{n s_x s_y},$$

where s_x and s_y are the square roots of the corresponding variance estimates. This estimate defines what is commonly called the *sample correlation coefficient*.

2.7. Confidence intervals

A difficulty with the methods of estimation presented thus far is that they do not tell us how close to the parameter being estimated the estimate is likely to be. We shall now study a method of estimation that determines the accuracy of an estimator of a single parameter θ. This method is most easily described by means of a particular example; therefore consider the problem of estimating the mean of a normal density with given variance σ^2 by means of a sample of size n. We know that $\bar{X}$ is a best unbiased estimator of the mean; consequently we shall base our approach on it. Since $\bar{X}$ possesses the density

$$(19) \qquad f(\bar{x} \mid \theta) = \frac{\sqrt{n} \exp \left[-n(\bar{x} - \theta)^2/2\sigma^2\right]}{\sqrt{2\pi}\sigma},$$

we can calculate the probability that $\bar{X}$ will lie within any specified interval about θ. In this connection it follows from Table IV in the appendix that if Z is a standard normal variable

$$P(|Z| < 2) = .95.$$

It will be observed from (19) that $\dfrac{(\bar{X} - \theta)\sqrt{n}}{\sigma}$ is a standard normal variable; consequently we may write

$$P\left(\frac{|\bar{X} - \theta|\sqrt{n}}{\sigma} < 2\right) = .95.$$

This is equivalent to

$$P\left(-2 < \frac{(\bar{X} - \theta)\sqrt{n}}{\sigma} < 2\right) = .95.$$

By solving these inequalities for θ, we may rewrite our probability in the form

$$P\left(\bar{X} - \frac{2\sigma}{\sqrt{n}} < \theta < \bar{X} + \frac{2\sigma}{\sqrt{n}}\right) = .95.$$

If this probability is interpreted operationally in terms of the relative frequency with which the indicated event will occur in the long run of similar sampling experiments, the preceding probability states that in the long run 95 percent of the intervals of the form $(\bar{X} - 2\sigma/\sqrt{n}, \bar{X} + 2\sigma/\sqrt{n})$, which are based on samples of n each, will contain the mean of the normal density in their interior. Since $\bar{X}$ is a random variable and determines the midpoint of the interval, this interval shifts its location from sample to sample. The important property of these intervals is that in the long run 95 percent of them will contain the fixed point θ in their interior. Since all of these intervals have length $4\sigma/\sqrt{n}$, we obtain not only the point estimate $\bar{X}$ of θ but also a measure of the accuracy of our estimate in the sense that we should be willing to give odds of 19 to 1 in a wager that θ, although unknown to us, will be found to lie inside our interval. The interval $(\bar{X} - 2\sigma/\sqrt{n}, \bar{X} + 2\sigma/\sqrt{n})$ is called a 95 *percent confidence interval* for θ, and the end points of the interval are called 95 *percent confidence limits* for θ.

Now suppose that a sample of size 25 gave the value $\overline{X} = 20$ and that $\sigma = 1$. Substituting these values into our formula will produce the interval $(19.6, 20.4)$. We do not make a probability statement about this particular interval as we did for the random interval $(\overline{X} - 2\sigma/\sqrt{n}, \overline{X} + 2\sigma/\sqrt{n})$; nevertheless it is customary to call a numerical interval obtained in this manner a 95 percent confidence interval. We have 95 percent confidence in our numerical interval containing θ in the sense that in the long run 95 percent of the numerical intervals constructed in this manner will contain θ in their interior.

The preceding confidence interval for the mean of a normal variable was constructed on the assumption that the value of σ was known. A different technique, which will be presented in the next chapter, is needed if the value of σ is unavailable.

As a second illustration of the method of confidence intervals, consider the problem of finding a 90 percent confidence interval for the parameter σ^2 of a normal density with known mean μ. We know from Section 2.1 that $\sum (X_i - \mu)^2/n$ is a best unbiased estimator of σ^2; therefore we shall use it to construct a confidence interval for σ^2. We also know from an earlier discussion that $\sum (X_i - \mu)^2/\sigma^2$ possesses a χ^2 distribution with n degrees of freedom; therefore from Table V in the appendix we can find two numbers χ_1^2 and χ_2^2 satisfying

$$P(\chi^2 < \chi_1^2) = .05 \qquad \text{and} \qquad P(\chi^2 > \chi_2^2) = .05.$$

Hence,

$$P(\chi_1^2 < \chi^2 < \chi_2^2) = .90$$

and

(20)
$$P\left(\chi_1^2 < \frac{\sum (X_i - \mu)^2}{\sigma^2} < \chi_2^2\right) = .90.$$

By solving these inequalities for σ^2, this probability may be rewritten in the form

$$P\left(\frac{\sum (X_i - \mu)^2}{\chi_2^2} < \sigma^2 < \frac{\sum (X_i - \mu)^2}{\chi_1^2}\right) = .90.$$

The desired 90 percent confidence interval for σ^2 is therefore the interval whose end points are

$$\frac{\sum (X_i - \mu)^2}{\chi_2^2} \qquad \text{and} \qquad \frac{\sum (X_i - \mu)^2}{\chi_1^2}.$$

In Section 2.5.1 use was made of the fact that the variable

$$\frac{\sum_{i=1}^{n} (X_i - X)^2}{\sigma^2}$$

possesses a chi-square distribution with $n - 1$ degrees of freedom. If we do not know the value of μ, it suffices to replace $\sum (X_i - \mu)^2/\sigma^2$ by

$\sum (X_i - \overline{X})^2/\sigma^2$ in (20) and proceed as before. Now, however, the values of χ_1^2 and χ_2^2 will be the table values based on $n - 1$ degrees of freedom.

The preceding examples illustrate a widely applicable method for finding confidence intervals. It consists in first finding some random variable involving a sample statistic and the parameter θ whose distribution does not depend upon any parameters. An interval type of probability statement is then made about this random variable. If this statement can be rearranged to yield an interval statement concerning θ, the interval that results is the desired confidence interval.

The construction of the confidence intervals discussed in the preceding paragraphs was based upon the assumption that X possesses a normal distribution. A natural question to ask in this connection is, "How accurate are those confidence intervals if X does not possess a normal distribution?"

Investigations have shown that the confidence interval for μ based on $\overline{X}$ is highly reliable even though X possesses a distribution that differs considerably from normality. This is closely related to the Central Limit Theorem as it applies to $\overline{X}$. Other investigations, however, have shown that the confidence interval for σ^2 based on treating $\sum (X_i - \overline{X})^2/\sigma^2$ as a χ^2 variable is quite inaccurate if, for example, X has a distribution whose fourth moment is large compared to that for a normal variable having variance σ^2. The confidence interval based on a normality assumption may then be considerably shorter than it should be. Thus, it is usually safe to use the preceding methods for finding a confidence interval for μ even though the distribution of X is not necessarily normal; however the technique for finding a confidence interval for σ^2 should not be employed unless it is quite certain that X possesses a normal distribution, at least to a good approximation. Problems of the preceding type that arise from a lack of theoretical assumptions being satisfied will be treated in more detail in the last chapter.

BAYESIAN METHODS

2.8. Bayesian estimation

We shall now consider how to modify estimation methods when the parameter θ may be treated as a random variable. We assume as before that a random sample $X_1, \ldots, X_n$ is to be taken and to be used for estimating θ. Previously $f(x \mid \theta)$ denoted the probability density of the random variable X with θ being a fixed constant. Now that θ is a random variable, $f(x \mid \theta)$ represents the conditional probability density of X, given that θ is held fixed. Although θ is being treated as a random variable in the general framework, the sample $X_1, \ldots, X_n$ is to be used to estimate the particular value of θ that applies to our experiment. When dealing with probability densities we shall denote the joint density of our two random variables by $f(x, \theta)$. Then, if θ possesses the density $\pi(\theta)$, it follows from the definition of a conditional density that

$$f(x, \theta) = \pi(\theta)f(x \mid \theta).$$

Under random sampling, the joint distribution of all the random variables will be given by the density

(21) $$f(x_1, \ldots, x_n, \theta) = \pi(\theta) \prod_{i=1}^{n} f(x_i \mid \theta).$$

To simplify the notation, we shall use the device introduced in Chapter 1 of representing multiple integration with respect to the x's by means of a single x integral. As a result, if we choose the same loss function as before, namely squared error, and assume that we are dealing with continuous random variables, Formulas (2) and (4) of Chapter 1 will give the mean risk

$$r(\pi, d) = \int \int (d(x) - \theta)^2 f(x \mid \theta) \pi(\theta) \, dx \, d\theta.$$

The use of (21) will change this to the form

(22) $$r(\pi, d) = \int \int (d(x) - \theta)^2 f(x, \theta) \, dx \, d\theta.$$

In this condensed notation $f(x, \theta)$ represents (21) and dx represents $dx_1 \ldots dx_n$.

Now let the joint density $f(x, \theta)$ be expressed in the form

(23) $$f(x, \theta) = g(x)h(\theta \mid x),$$

where $g(x)$ denotes the marginal density of the random variable $X = (X_1, \ldots, X_n)$ and $h(\theta \mid x)$ denotes the conditional density of θ, given that $X = x$. This conditional density defines the *posterior probability distribution of θ, given the sample*. Thus, the prior density $\pi(\theta)$ gives the distribution of θ before sampling, whereas the posterior density $h(\theta \mid x)$ gives the distribution of θ after sampling. If (23) is used in (22) and if the order of integration is changed in (22), then (22) will assume the form

(24) $$r(\pi, d) = \int g(x) \left[\int (\theta - d(x))^2 h(\theta \mid x) \, d\theta \right] dx.$$

Since all quantities being integrated are nonnegative, it follows that $r(\pi, d)$ will be minimized if the quantity in brackets, which is a function of x, is minimized for each value of x. This in turn will be realized if the decision function $d(x)$ is chosen as that function of x which minimizes the integral in brackets for each value of x. But since $h(\theta \mid x)$ is the conditional density of θ given x, this integral is merely the second moment of that distribution about the point $d(x)$. From Volume I we know that the second moment is a minimum when it is taken about the mean of the distribution; therefore this integral will be minimized when $d(x)$ is chosen to be the mean of the conditional distribution represented by $h(\theta \mid x)$. Under the assumptions that produced (24) these arguments constitute a proof of the following theorem.

Theorem 2   A Bayes solution for the estimation problem is given by choosing the decision function

$$d(x) = E(\theta \mid x)$$

that is, by choosing $d(x)$ to be the mean of the posterior distribution of θ.

As an illustration, consider the problem of estimating the parameter p of a Bernoulli distribution on the basis of a sample of size n, given that p possesses a beta prior density given by

$$(25) \qquad \pi(p) = \frac{\Gamma(\alpha + \beta)}{\Gamma(\alpha)\Gamma(\beta)} p^{\alpha - 1}(1 - p)^{\beta - 1}, \qquad 0 < p < 1.$$

Writing out $f(x, p)$ in its uncondensed form as given in (21),

$$(26) \qquad f(x, p) = \frac{\Gamma(\alpha + \beta)}{\Gamma(\alpha)\Gamma(\beta)} p^{\alpha - 1}(1 - p)^{\beta - 1} p^{\Sigma x_i}(1 - p)^{n - \Sigma x_i}$$

$$= \frac{\Gamma(\alpha + \beta)}{\Gamma(\alpha)\Gamma(\beta)} p^{\alpha + S - 1}(1 - p)^{\beta + n - S - 1}$$

where $S = \sum_{i=1}^{n} x_i$ represents the total number of successes in n trials. Now

$$E(p \mid x) = \int_0^1 p h(p \mid x) \, dp$$

$$= \int_0^1 p \frac{f(x, p)}{g(x)} \, dp$$

$$= \frac{\int_0^1 p f(x, p) \, dp}{\int_0^1 f(x, p) \, dp}.$$

Application of (26) will give

$$E(p \mid x) = \frac{\int_0^1 p^{\alpha + S}(1 - p)^{\beta + n - S - 1} \, dp}{\int_0^1 p^{\alpha + S - 1}(1 - p)^{\beta + n - S - 1} \, dp}.$$

Since the integral of $\pi(p)$ over $[0, 1]$ is 1, it follows from (25) that

$$\int_0^1 p^{r - 1}(1 - p)^{s - 1} \, dp = \frac{\Gamma(r)\Gamma(s)}{\Gamma(r + s)}.$$

This formula together with the property of the gamma function that $\Gamma(r + 1) = r\Gamma(r)$ will yield the result

$$(27) \quad E(p \mid x) = \frac{\Gamma(\alpha + S + 1)\Gamma(\beta + n - S)}{\Gamma(\alpha + \beta + n + 1)} \cdot \frac{\Gamma(\alpha + \beta + n)}{\Gamma(\alpha + S)\Gamma(\beta + n - S)}$$

$$= \frac{\alpha + S}{\alpha + \beta + n} \cdot$$

As a special case, let us choose $\alpha = \beta = 1$. The beta density given by (25) then reduces to the uniform density over $[0, 1]$. Then

$$E(p \mid x) = \frac{S + 1}{n + 2} \cdot$$

Thus, the Bayesian estimate for p differs from our previous best unbiased non-Bayesian estimate, namely the success ratio S/n, by requiring that an additional success be added to S and two additional trials be added to n. For large values of n there is very little difference in these estimates; however, suppose only one trial is made. Then the classical estimate gives 0 or 1 as the estimate of p, depending upon whether a failure or a success occurred. The corresponding Bayesian estimates are $1/3$ or $2/3$. If p were the probability of getting a head in tossing a coin selected from one's purse, the Bayes estimate would be much more realistic because our knowledge of coins would tend to reject an estimate of 0 or 1. Such knowledge would also suggest using a more realistic prior density than the uniform one. For example, the choice of $\alpha = \beta = 5$ would yield a prior distribution with fairly high probability density at $p = 1/2$ and low density near 0 and 1, which is more in keeping with our knowledge of coins. For a single toss of the coin, Formula (27) would then yield the estimates $5/11$ or $6/11$.

As a second illustration, consider the problem of estimating the mean μ of a normal density if the value of σ is known, and if the prior density is also normal. Let the latter density be given by

$$\pi(\mu) = \frac{\exp\left[-\frac{1}{2}\left(\frac{\mu - \mu_0}{\alpha}\right)^2\right]}{\sqrt{2\pi\alpha}} \cdot$$

If our estimate is to be based on a random sample of size n, the joint density is

$$f(x, \mu) = \frac{\exp\left[-\frac{1}{2}\left(\frac{\mu - \mu_0}{\alpha}\right)^2\right]}{\sqrt{2\pi\alpha}} \cdot \frac{\exp\left[-\frac{1}{2}\sum_{i=1}^{n}\left(\frac{x_i - \mu}{\sigma}\right)^2\right]}{(2\pi)^{n/2}\sigma^n} \cdot$$

Rather than calculate $E(\mu \mid x)$ directly, we shall first calculate $h(\mu \mid x)$ in order to have it available for another investigation. To calculate $h(\mu \mid x)$

we must calculate the marginal density $g(x)$ which is obtained by integrating $f(x, \mu)$ with respect to μ. This is accomplished by: squaring out the two binomials in the exponent of $f(x, \mu)$, collecting terms in μ^2 and μ, completing the square in μ, and then performing the integration. These algebraic manipulations will give

$$\left(\frac{\mu - \mu_0}{\alpha}\right)^2 + \sum_{i=1}^{n} \left(\frac{x_i - \mu}{\sigma}\right)^2 = s(\mu) + t,$$

where t does not depend upon μ and where

$$s(\mu) = \frac{\sigma^2 + \alpha^2 n}{\alpha^2 \sigma^2} \left(\mu - \frac{\sigma^2 \mu_0 + \alpha^2 n \bar{x}}{\sigma^2 + \alpha^2 n}\right)^2.$$

As a result we may write $f(x, \mu)$ in the form

(28) $$f(x, \mu) = ce^{-(1/2)[s(\mu) + t]}.$$

Since

$$h(\mu \mid x) = \frac{f(x, \mu)}{\int_{-\infty}^{\infty} f(x, \mu) \, d\mu},$$

it follows from (28) that c and the term in t will cancel in this ratio and hence that

$$h(\mu \mid x) = \frac{e^{-(1/2)s(\mu)}}{\int_{-\infty}^{\infty} e^{-(1/2)s(\mu)} \, d\mu}.$$

Now except for the proper constant, the form of $s(\mu)$ shows that $e^{-(1/2)s(\mu)}$ is the density of a normal variable μ; therefore, if the numerator and denominator are multiplied by this constant, the integral in the denominator will be equal to one. The proper constant is the reciprocal of $\sqrt{2\pi}$ times the standard deviation of the normal variable. The latter can be obtained by inspecting $s(\mu)$, hence

$$h(\mu \mid x) = \frac{\exp\left[-\dfrac{\sigma^2 + \alpha^2 n}{2\alpha^2 \sigma^2}\left(\mu - \dfrac{\sigma^2 \mu_0 + \alpha^2 n \bar{x}}{\sigma^2 + \alpha^2 n}\right)^2\right]}{\sqrt{\dfrac{2\pi\alpha^2\sigma^2}{\sigma^2 + \alpha^2 n}}}.$$

Since the Bayesian estimate is the mean of this normal variable, our estimate is given by the formula

$$E(\mu \mid x) = \frac{\sigma^2 \mu_0 + \alpha^2 n \bar{x}}{\sigma^2 + \alpha^2 n}.$$

An equivalent but more useful version is the formula

$$E(\mu \mid x) = \frac{\delta}{1 + \delta} \mu_0 + \frac{1}{1 + \delta} \bar{x},$$

where $\delta = \sigma^2/n\alpha^2$ is the ratio of the variance of the sample mean $\bar{X}$ and the variance of the prior density $\pi(\mu)$.

If the variance α^2 of the prior density becomes infinite, which implies that there is no preference for one value of μ over any other, the Bayesian estimate will approach the classical non-Bayesian estimate $\bar{X}$. However, if the variance α^2 is very small compared to the variance of $\bar{X}$, the value of δ will be large and the Bayesian estimate will be close to μ_0. In that case the data estimate $\bar{X}$ will contribute very little information to the problem. For fixed values of α^2 and σ^2, the Bayesian estimate will approach the classical estimate as $n \rightarrow \infty$, which means that the data information is swamping out the information contributed by the prior density.

For the purpose of comparing the classical estimate $\bar{X}$ with the Bayesian estimate, we should compare the mean risk for both estimates. For squared error loss the risk function for $\bar{X}$ is equal to its variance, consequently $\mathcal{R}(\theta, \bar{x}) = \sigma^2/n$, and therefore $r(\pi, \bar{x}) = \sigma^2/n$. From (24) it will be observed that the integral in brackets is the variance of the conditional density $h(\mu \mid x)$ when the Bayes estimate is used. Hence by inspecting $h(\mu \mid x)$ we find that this variance is $\dfrac{\alpha^2\sigma^2}{\sigma^2 + \alpha^2 n}$. Integration of this constant with respect to the marginal density $g(x)$ then gives the Bayes risk $r(\pi, E(\mu \mid x)) = \dfrac{\alpha^2\sigma^2}{\sigma^2 + \alpha^2 n}$. Using the ratio of these mean risk values for making the comparison, we obtain the result

$$\frac{r(\pi, E(\mu \mid x))}{r(\pi, \bar{x})} = \frac{1}{1 + \delta} .$$

For a situation in which $\pi(\mu)$ is available, this formula makes it a simple matter to check on the relative importance of using the prior density $\pi(\mu)$ as contrasted to ignoring it.

To indicate the importance of using prior information when it is available, suppose we have the following problem. A manufacturer of steel rods sends a shipment each month to a firm that uses them in its product. This firm is interested in the tensile strength of the rods and tests samples for this property at regular intervals. Suppose that over a long period of time it has found that the mean tensile strength μ of a shipment may be treated as a normal variable with a long run average of $\mu_0 = 1,000$ and a standard deviation for the shipment mean of $\alpha = 5$. Further, suppose that tests made on individual rods of a shipment show that the tensile strength X of an individual rod may be treated as a normal variable with mean μ and standard deviation $\sigma = 50$. Now if a fresh shipment comes in and a sample of $n = 25$ is tested, compare the estimate $\bar{X}$ for that particular shipment mean μ with the corresponding Bayesian estimate.

For this problem $\delta = \dfrac{(50)^2}{25(5)^2} = 4$. The Bayes estimate is therefore

$$E(\mu \mid x) = (4/5)(1,000) + (1/5)\overline{X}$$

and the ratio of the mean risks is 1/5. Thus, there is a large advantage in using the Bayesian estimate.

A very useful property of Bayesian methods is that additional experimental data can be conveniently incorporated into a revised posterior distribution. If two successive sampling experiments are to be made, it is possible after the first experiment to calculate the posterior distribution, and treat it as a prior distribution for the second experiment in the calculation of the posterior distribution after both experiments. This property will not be demonstrated here.

2.9. Prediction problems

A problem closely related to that studied in the preceding section is that of trying to predict the value of one random variable by means of the known value of a related random variable. Suppose we have two random variables X and Y that possess the density $f(x, y)$. Suppose further that we wish to estimate the value of the variable Y on the basis of having been told the value of the variable X. For example, X and Y might represent the height of a father and son, and we would like to predict the height of a son from a knowledge of his father's height. If we choose as our predictor the function $d(x)$ and if we use squared error as our criterion for determining how good an approximation the predictor $d(x)$ gives, then we should try to choose $d(x)$ to be that function which will minimize

(29) $$E(Y - d(x))^2 = \int (y - d(x))^2 f(y \mid x) \, dy.$$

This assumes that we are given a particular value of the random variable X and wish to predict the associated value of Y. If $d(x)$ is chosen to be that function of x which minimizes this integral for each value of x, the predictor $d(x)$ will be optimal for every x that may arise.

A comparison of the integral in brackets in (24) with the integral in (29) shows that the two minimization problems are the same except for notation; consequently, the best predictor is given by the function

$$d(x) = E(Y \mid x).$$

Although this predictor was derived on the basis of knowing the value of X, it will be optimal if X is chosen at random because it is optimal for every possible value of X; consequently this predictor will minimize

(30) $E(Y - d(X))^2 = \int \int (y - d(x))^2 f(y \mid x) g(x) \, dy \, dx$

$$= \int \int (y - d(x))^2 f(x, y) \, dy \, dx.$$

Here $g(x)$ denotes the marginal density of X and the left side is the expectation with respect to both X and Y and not just Y, as was the case in (29).

Just as in the preceding section, these various formulas are valid whether X is a single random variable or represents a vector variable $X = (X_1, \ldots, X_n)$.

Problems of prediction are very common. For example, if the components of X represent the score a student makes on a college aptitude test, the score he makes on a placement test, and his high school grade point average, then a college will try to predict his college grade point average Y by means of the value of X.

The calculation of $E(Y \mid x)$ requires knowledge of the conditional density $f(y \mid x)$ or of the joint density $f(x, y)$. Unfortunately, it is often difficult to postulate a precise density of this type when several variables are involved. Unless there is a large amount of earlier sampling experience available or good prior theoretical information, it is difficult to determine the nature of a multidimensional distribution. Under such circumstances one often resorts to linear prediction since such prediction does not require complete knowledge of the distribution. Thus, we choose a predictor of Y corresponding to an X to be of the form

$$d(X) = a + bX.$$

The problem of optimal prediction then is the problem of choosing a and b to minimize

$$E(Y - a - bX)^2.$$

Here, as in (30), expectation is with respect to both variables. The formulas that result from this minimization become simpler if X is measured from its mean μ_x, therefore we shall reformulate the problem to that of trying to determine the values of a and b that will minimize

$$E[Y - a - b(X - \mu_x)]^2.$$

The computations needed to evaluate this expectation are simplified by introducing the mean μ_y and using familiar properties of the expectation operator that were demonstrated in Volume I. Thus, we calculate

$$E[Y - a - b(X - \mu_x)]^2 = E[(Y - \mu_y) - b(X - \mu_x) + (\mu_y - a)]^2$$

$$= E[(Y - \mu_y)^2 + b^2(X - \mu_x)^2 + (\mu_y - a)^2$$

$$- 2b(X - \mu_x)(Y - \mu_y) + 2(Y - \mu_y)(\mu_y - a) - 2b(X - \mu_x)(\mu_y - a)].$$

The expectation of the last two terms is zero, therefore it follows from the definitions of the variance and correlation coefficient that

$$E[Y - a - b(X - \mu_x)]^2 = \sigma_y^2 + b^2\sigma_x^2 + (\mu_y - a)^2 - 2b\rho\sigma_x\sigma_y.$$

For a calculus relative minimum it is necessary that the partial derivatives with respect to a and b vanish; therefore, a and b must satisfy the equations

$$-2(\mu_y - a) = 0$$

and

$$2b\sigma_x^2 - 2\rho\sigma_x\sigma_y = 0.$$

The solution of these two equations is given by

$$a = \mu_y \quad \text{and} \quad b = \rho\frac{\sigma_y}{\sigma_x}.$$

This unique solution can be seen to yield both a calculus relative minimum and an absolute minimum, hence our best predictor is the function

(31)
$$d(x) = \mu_y + \rho\frac{\sigma_y}{\sigma_x}(x - \mu_x).$$

This predictor requires only the knowledge of the first two moments and the correlation coefficient of the random variables X and Y. Since such quantities can often be estimated with precision if they are not available, it is a highly useful type of predictor. Of course, if there is reason to believe that $E(Y \mid x)$ is far from being linear in x, a linear predictor would be of little value. If the sample correlation r has a value close to 1, there is good reason to use a linear predictor because we know that $\rho = 1$ implies that there exists a linear relationship between X and Y. It is possible, however, for a linear predictor to be the best predictor even though the value of ρ is small, therefore a small value of r does not imply that $E(Y \mid x)$ cannot be approximated satisfactorily by a linear function. On the other hand, if ρ is small it is possible for there to be an excellent nonlinear predictor. These possibilities will be discussed further in Chapter 4.

One measure of how good the linear predictor given by (31) is for prediction purposes can be obtained by comparing the expected squared error when (31) is used to that when the value of X is ignored. Since $c = \mu_y$ minimizes $E(Y - c)^2$, this implies that the two quantities to be compared are

$$E(Y - d(X))^2 \quad \text{and} \quad E(Y - \mu_y)^2 = \sigma_y^2.$$

Writing (31) in the form $d(X) = \mu_y + b(X - \mu_x)$, we can express the first of these expected values as

$$E[Y - \mu_y - b(X - \mu_x)]^2 = E(Y - \mu_y)^2 - 2bE(X - \mu_x)(Y - \mu_y)$$

$$+ b^2 E(X - \mu_x)^2$$

$$= \sigma_y^2 - 2b\rho\sigma_x\sigma_y + b^2\sigma_x^2.$$

Substituting the value of b and simplifying will yield the result

$$E(Y - d(X))^2 = \sigma_y^2(1 - \rho^2).$$

Thus, the ratio of the mean squared error that arises when the best linear predictor is used to the mean squared error when X is ignored equals $1 - \rho^2$.

As an illustration, it is known that for certain classes of schools the correlation between high school and college grade point averages is in the neighborhood of .5; consequently, the mean squared error that will occur when predicting a student's college grade point average by a linear predictor based on his high school grade point average will be 3/4ths as large as that occurring if the high school record is ignored. A few years ago a certain university studied the relationship between a student's college aptitude test score and his freshman grade point average and found a correlation of .10. This implies that the mean squared error of predicting a student's college record by his aptitude test score is only 1 percent lower than that obtained by predicting that his record will be the mean record for all freshmen. This university naturally ceased using that particular test for forecasting future performance. It takes a relatively large value of ρ to make $1 - \rho^2$ small; therefore, a linear predictor is not likely to be very useful unless the correlation coefficient is fairly large.

Exercises

1 Given $f(x \mid \theta) = \dfrac{e^{-x^2/2\theta}}{\sqrt{2\pi\theta}}$, show that $d(X) = \dfrac{\sum_{i=1}^{n} X_i^2}{n}$, where $X_1, \ldots, X_n$ is a random sample from $f(x \mid \theta)$, is an unbiased estimator of θ.

2 Given $f(x \mid \theta) = 1/\theta, 0 \le x \le \theta$, determine c so that $d(X) = cX$ will be an unbiased estimator of θ.

3 Given $f(x \mid \theta) = \dfrac{x^{\theta-1}e^{-x}}{\Gamma(\theta)}$, $x > 0, \theta > 0$,

   (a) find a value of c such that $d(X) = cX$ will be an unbiased estimator of θ;

   (b) determine whether it is possible to find a value of c such that $d(X) = cX^2$ will be an unbiased estimator of θ.

4 Determine the restrictions that are needed on the a's so that $Z = \sum_{i=1}^{n} a_i X_i$ will be an unbiased estimator of EX, where $X_1, \ldots, X_n$ represents a random sample of X.

5 If the bias of an estimator is denoted by $b(\theta)$ and is defined by the relation $Ed(X) = \theta + b(\theta)$, calculate the bias in the minimax estimator $d'(X) = (X + \sqrt{n}/2)/(n + \sqrt{n})$ associated with Figure 1, where X is a binomial variable with parameters n and p.

6 Show that the value of $\mathscr{R}(p, d') = E(d' - p)^2$, where d' is the minimax estimator of Exercise 5, is given by $1/(4(\sqrt{n} + 1)^2)$.

7 Fish are caught in a lake until n fish of a certain type have been obtained. Let X denote the total number of fish caught at that point. Assume that the lake is so large that the extraction of these fish does not change the proportion p of this type of fish in the lake. Show that the random variable X possesses the density function given by $\binom{x - 1}{n - 1} p^n (1 - p)^{x-n}$, $x = n, n + 1, \ldots$, where n is assumed to be a fixed integer. Use this result to show that $(n - 1)/(X - 1)$ is an unbiased estimator of p. Note that this estimator differs from the intuitive estimator n/X, which is biased for this problem.

8 Show that $d = \sum_{i=1}^{n} X_i/n$ is an efficient unbiased estimator of the parameter θ for the Poisson density $f(x \mid \theta) = e^{-\theta}\theta^x/x!$, $x = 0, 1, \ldots$, where $X_1, \ldots, X_n$ represents a random sample of X.

9 (a) What is the best set of a's to choose in Exercise 4 if $Z = \sum_{i=1}^{n} a_i X_i$ is to be an unbiased estimator of EX with minimum variance?
   (b) What a's should be chosen if the X_i are independent variables with the same mean but different known variances σ_i^2?

10 Show that the estimator $Z = (X_1 + X_2)/2$ is an unbiased estimator of EX, where $X_1, \ldots, X_n$ represents a random sample of X. Calculate the efficiency of Z if X possesses the density

$$f(x \mid \theta) = \frac{\exp\left[-(1/2)(x - \theta)^2\right]}{\sqrt{2\pi}}.$$

11 Let X be a random variable with finite second moment. Show that $E(X - b)^2$, where b is a constant, is minimized by $b = EX$.

12 Find the lower bound of the variance for unbiased estimators of the parameter θ for the Cauchy density $f(x \mid \theta) = \dfrac{1}{\pi[1 + (x - \theta)^2]}$.

13 Why is the derivation that produced the inequality in Theorem 1 not applicable to the problem of estimating θ for the uniform density $f(x \mid \theta) = 1/\theta$, $0 \le x \le \theta$?

14 Calculate the mean and variance of the estimator $Z = \dfrac{\sum_{i=1}^{n} (X_i - \mu)^2}{n}$

for the parameter θ in $f(x \mid \theta) = \dfrac{e^{-[(x-\mu)^2/2\theta]}}{\sqrt{2\pi\theta}}$. Use the fact from

Volume I that $(X - \mu)^2/\theta$ possesses a chi-square distribution with one degree of freedom and hence that nZ/θ possesses a chi-square distribution with n degrees of freedom. This assumes that $X_1, \ldots, X_n$ represent a random sample of X.

15 Use the result of Exercise 14 to show that Z as defined there is an efficient unbiased estimator of θ.

16 By differentiating (5) under the integral sign with respect to θ and using the relationship

$$\frac{\partial L}{\partial \theta} = L \frac{\partial \log L}{\partial \theta},$$

show that

$$E\left(\frac{\partial \log f(X \mid \theta)}{\partial \theta}\right)^2 = -E\left(\frac{\partial^2 \log f(X \mid \theta)}{\partial \theta^2}\right).$$

This gives an alternative method for calculating the lower bound in (2) when it is applicable.

17 Find the maximum likelihood estimator of θ for $f(x \mid \theta) = \theta e^{-\theta x}$, $x > 0$, based on a random sample of size n.

18 Given the information of Exercise 17, find the maximum likelihood estimator of EX and compare it with the result obtained in Exercise 17.

19 Find the maximum likelihood estimator of p for the geometric density $f(x \mid p) = p(1 - p)^x$, $x = 0, 1, 2, \ldots$, based on a random sample of size n.

20 Perform the following experiment twenty times. Roll a die until a four spot is obtained and record the number of rolls that were required. Let $x_1, \ldots, x_{20}$ represent the experimental outcomes in 20 such experiments. Use these values and the result in Exercise 19 to estimate the value of p, which is 1/6, for this problem.

21 Find the maximum likelihood estimator of θ, based on a random sample of size n, for the density $f(x \mid \theta) = (1 + \theta)x^\theta$, $0 \le x \le 1$, $\theta > -1$.

22 Given $f(x \mid \theta) = \dfrac{e^{-(x^2/2\theta^2)}}{\sqrt{2\pi\theta}}$, find the maximum likelihood estimator of θ based on a random sample of size n.

23 Treat $\varphi = \theta^2$ as the parameter in Exercise 22 and then find the maximum likelihood estimator of φ. Compare your result with the result obtained in Exercise 22.

24 Given the following random sample values of a normal variable, for which $\mu = 10$ and $\sigma = 3$, use them and (10) to obtain estimates of μ and σ^2: 12.7, 6.6, 5.6, 14.3, 11.4, 4.3, 7.2, 10.8, 13.8, 11.2, 10.0, 12.8, 7.1, 14.0, 6.1.

25 Use Table III of the appendix to obtain a random sample of size 10 of a standard normal variable. Use your sample values and (10) to obtain estimates of μ and σ, which are 0 and 1 here.

26 Use the second derivative test to show that $\hat{p} = \sum_{i=1}^{n} X_i/n$ does maximize the likelihood function for the Bernoulli density discussed in Section 2.4.

27 Use the results of Exercise 15 to show that the maximum likelihood estimator of Exercise 23 is an efficient unbiased estimator.

28 Given $f(x \mid \theta) = 1/\theta, 0 \leq x \leq \theta$, show that for a random sample of size n the estimator $\hat{\theta}$ that maximizes the likelihood function $(1/\theta)^n$ is given by $\hat{\theta} = \max (X_1, \ldots, X_n)$, that is, the largest of the sample values. This is an illustration of a problem for which the usual calculus techniques of finding a maximizing value are inappropriate.

29 Find the value of k that minimizes $E(ks^2 - \sigma^2)^2$, where

$$s^2 = \frac{\sum (X_i - \mu)^2}{n},$$

and the X_i represent a random sample of a normal variable. Use the fact that $\frac{\sum (X_i - \mu)^2}{\sigma^2}$ is a chi-square variable with n degrees of freedom and that the mean and variance of such a variable are equal to n and $2n$, respectively. What does this result show about the requirement of unbiasedness if expected squared error is the basis for comparing estimators, rather than the variance?

30 Use the method of moments to estimate the parameter θ for the uniform density $f(x \mid \theta) = 1/\theta, 0 \leq x \leq \theta$, based on a random sample of size n.

31 Use Table II of the appendix to obtain a random sample of size 20 from the uniform density $f(x \mid \theta) = 1/\theta, 0 \leq x \leq \theta$, where $\theta = 1$. Use your sample values and the result in Exercise 30 to estimate the value of θ.

32 Use the method of moments to estimate the parameter θ for the density
$$f(x \mid \theta) = \frac{x \exp [-(x^2/2\theta)]}{\theta}, \quad x > 0, \theta > 0, \text{ based on a random}$$
sample of size n and employing
(a) the first moment,
(b) the second moment.

33 Use the method of moments to estimate the parameters p and n for the binomial density $f(x \mid \theta) = \binom{n}{x} p^x(1 - p)^{n-x}$ based on the m random sample values $X_1, \ldots, X_m$.

34 Perform the following experiment twenty times. Toss a coin eight times and record the number of heads obtained. Let $x_1, \ldots, x_{20}$ represent the experimental outcomes. Use those values and the result in Exercise 33 to estimate the values of p and n, which are 1/2 and 8 here.

35 Show that the mean and variance of the gamma density

$$f(x \mid \alpha, \beta) = \frac{e^{-(x/\beta)}x^{\alpha-1}}{\beta^\alpha \Gamma(\alpha)}, \qquad x > 0, \quad \alpha > 0, \quad \beta > 0,$$

are given by the formulas $\mu = \alpha\beta$ and $\sigma^2 = \alpha\beta^2$.

36 Why would the method of moments fail in trying to estimate the parameter θ for the Cauchy density $f(x \mid \theta) = \dfrac{1}{\pi[1 + (x - \theta)^2]}$?

37 Show that the kth sample moment about the origin, based on a random sample of size n, is an unbiased estimate of the kth theoretical moment of a random variable X.

38 (a) How would you estimate $E(1/\sqrt{X})$, based on a random sample of size n, if you did not know the density of X?
   (b) Compare your estimate with the maximum likelihood estimate of $E(1/\sqrt{X})$ if X possesses the density $f(x \mid \theta) = e^{-(x/\theta)}/\theta$, $x > 0$.

39 Given $\overline{X} = 20$ for a random sample of size 25 from the density $f(x \mid \mu) = \dfrac{e^{-[(x-\mu)^2/32]}}{4\sqrt{2\pi}}$, find a 90% confidence interval for μ.

40 How large a sample would be needed in Exercise 39 if the resulting confidence interval were to be half as long?

41 Use Table III of the appendix to obtain a random sample of size 16 of a standard normal variable. Assume that you know $\sigma = 1$ but do not know that $\mu = 0$. Use your sample values to obtain an 80% confidence interval for μ.

42 Given $\sum X_i = 180$ and $\sum X_i^2 = 2{,}000$ for a random sample of size 20 from $f(x \mid \sigma) = \dfrac{\exp\left[-(1/2)[(x - 10)/\sigma]^2\right]}{\sigma\sqrt{2\pi}}$, find a 90% confidence interval for σ^2.

43 Use the sample values of Exercise 41 to obtain a 90% confidence interval for σ under the assumption that $\mu = 0$ but you do not know that $\sigma = 1$.

44 Work Exercise 43 under the assumption that μ is also unknown.

45 Given the following random sample values of a normal variable, use them to find a 95% confidence interval for σ^2: 6.70, 4.55, 7.12, 3.30, 2.15, 4.00, 3.38, 7.93, 8.40, 5.75, 2.91, 6.50, 7.96, 3.00, 4.93.

46 In estimating the mean of a normal variable by means of a confidence interval, how large a sample will be needed if the length of a 95% confidence interval is to be less than $\sigma/10$ if σ is assumed known?

47 Construct a 90% confidence interval for the parameter μ of the Poisson density if a sample of size 30 yielded $\sum X_i = 240$. Use the normal approximation for $\sum X_i/30$ and replace σ by its sample estimate.

48 Construct a 95% confidence interval for the parameter p of the Bernoulli density $f(x \mid p) = \begin{cases} p & \text{if } x = 1 \\ 1 - p & \text{if } x = 0 \end{cases}$ if a random sample of size 50 yielded $\sum X_i = 15$. Use the normal approximation for $\sum X_i/50$ and replace its variance by its sample estimate.

49 Toss a coin fifty times and record the number of heads obtained. Use this value and the result in Exercise 48 to obtain a 90% confidence interval for p, which is $1/2$ here.

50 Toss a die sixty times and record the number of six spots obtained. Use this value and the result in Exercise 48 to obtain an 80% confidence interval for p, which is $1/6$ here.

51 Construct an 80% confidence interval for the parameter θ of the density $f(x \mid \theta) = e^{-(x/\theta)}/\theta$, $x > 0$, if a random sample of size 20 yielded $\sum X_i = 80$. Use the normal approximation for $\sum X_i/20$.

52 Find the Bayes estimator of the Bernoulli parameter p based on a random sample of size n if $\mathscr{L}(p, a) = (a - p)^2$ and $\pi(p) = 3p(1 - p)$, $0 \le p \le 1$.

53 Given $f(x \mid \theta) = 1/\theta$, $0 \le x \le \theta$, $\mathscr{L}(\theta, a) = (a - \theta)^2$, and $\pi(\theta) = \theta e^{-\theta}$, $\theta > 0$,
(a) calculate the marginal density $f(x)$,
(b) use $f(x)$ to find the posterior density $f(\theta \mid x)$,
(c) use the result obtained in (b) to calculate $E(\theta \mid x)$.

54 Find the Bayes estimator of the Poisson parameter μ based on a random sample of size n if $\mathscr{L}(\mu, a) = (a - \mu)^2$ and $\pi(\mu) = e^{-\mu}$, $\mu > 0$.

55 Find the Bayes estimator of the parameter θ for the density $f(x \mid \theta) = \theta e^{-\theta x}$, $x > 0$, based on a single observed value of X, if

$$\mathscr{L}(\theta, a) = (a - \theta)^2 \quad \text{and} \quad \pi(\theta) = 1, \quad 0 < \theta < 1.$$

56 Given $f(x \mid \theta) = e^{-(x/\theta)}/\theta$, $x > 0$, $\pi(\theta) = (2/3)\theta$, $1 \le \theta \le 2$, and $\mathscr{L}(\theta, a) = (a - \theta)^2$,

(a) calculate $\mathcal{R}(\theta, d)$ for $d_1(X) = X$ and $d_2(X) = X - 1$, and determine which is the minimax estimator with respect to these two estimators,

(b) calculate the Bayes solution with respect to these two estimators.

57 Given $f(x \mid \theta) = 1/\theta$, $0 \le x \le \theta$, $\pi(\theta) = 1$, $0 < \theta < 1$, and $\mathcal{L}(\theta, a) = |a - \theta|$,

(a) calculate $\mathcal{R}(\theta, d)$ and $r(\pi, d)$ for $d(X) = X$;

(b) calculate $\mathcal{R}(\theta, d)$ and $r(\pi, d)$ for $d(X) = X^2$ and compare with the results in (a).

58 Given $f(x \mid \theta) = \dfrac{x^{\alpha-1} e^{-\theta x} \theta^{\alpha}}{\Gamma(\alpha)}$, $x > 0$, squared error loss, and

$$\pi(\theta) = \dfrac{\theta^{a-1} e^{-\beta\theta} \beta^{a}}{\Gamma(a)},$$

where α, a, and β are given positive constants, calculate the Bayes estimator of θ based on a single observed value of X.

59 Given $f(x, y) = 8xy$, $0 \le x \le 1$, $0 \le y \le x$,

(a) find the equation of the best predictor $d(x) = E(Y \mid x)$;

(b) what can be said about the best linear predictor for this problem?

60 Given $f(x, y) = \dfrac{2x + 4y}{3}$, $0 \le x \le 1$, $0 \le y \le 1$,

(a) find the equation of the best predictor $E(Y \mid x)$;

(b) find the equation of the best linear predictor given by (31);

(c) graph these two predictors to see how well the linear predictor approximates the best predictor.

61 Suppose you performed an experiment m times and obtained s successes. Use this information to find the posterior density of p given s if the prior density of p is uniform. If the experiment is performed n additional times and S successes are obtained, what will the Bayes estimate become using squared error if the posterior density of p is treated as the prior density for the second set of experiments? How does this compare with combining the two experiments and using the uniform prior density?

3

Testing Hypotheses

This chapter is concerned with methods for solving the first two of the three basic problems of statistics, namely the testing of hypotheses and the multiple decision problem. The latter problem will be treated in the section on Bayesian methods. Because of its complexity it will be studied rather briefly; therefore we shall be concerned here principally with finding best methods for testing hypotheses.

Given a probability density $f(x \mid \theta)$ and a sample $x_1, \ldots, x_n$ from it, a typical problem of testing a hypothesis, that is, a problem for which there are only two possible actions available, is to decide by means of a decision function $d = d(x_1, \ldots, x_n)$ whether $\theta \le \theta_0$ or $\theta > \theta_0$, where θ_0 is some specified value. For example, suppose a new heat treatment is being tried for the purpose of attempting to increase the tensile strength of steel rods. If θ_0 represents the mean tensile strength of such rods under the old treatment, the testing problem might be one of deciding whether the new treatment has been harmful or beneficial with respect to mean tensile strength.

We shall begin the study of testing hypotheses by considering a somewhat simpler problem than the preceding one. It is the problem of deciding whether the value of the parameter θ is θ_0 or θ_1. As an example, suppose a distributor of fishing equipment has received a shipment of unlabeled nylon fishing line from a Japanese firm. If he purchases only two types of line from this firm, namely 8 pound and 10 pound test line, he would like to decide rather quickly which quality line was received. This problem can be treated as one of testing the hypothesis, which will be denoted by H_0, that the mean breaking strength of the line is $\theta = 8$ against the alternative hypothesis, which will be denoted by H_1, that the mean breaking strength is $\theta = 10$. Since experiments with testing the quality of industrial products show that it is probably safe to assume that a variable such as the breaking strength of nylon line possesses a normal distribution, we shall introduce a random variable X to represent the breaking strength of a standard testing length of nylon line, and we shall assume that it possesses a normal distribution with mean θ, where θ has either the value 8 or 10. If we take several standard lengths of the line and test them, we will accumulate a set of observed values of the random variable X with which to make a decision.

Since there are only two possible actions that can be taken in a testing problem, namely accept H_0 or accept H_1, a decision function $d = d(x_1, \ldots, x_n)$ must separate n dimensional sample space into two parts. Let A_0 denote the part that is associated with accepting H_0, and A_1 the remaining part associated with accepting H_1. This means that if a random sample of X yields a point $x = (x_1, \ldots, x_n)$ that lies in A_0, we accept the hypothesis $H_0 \colon \theta = \theta_0$, whereas if it lies in A_1, we accept the alternative hypothesis $H_1 \colon \theta = \theta_1$.

As a simple illustration of how the regions A_0 and A_1 can be chosen on an intuitive basis, consider the following discrete variable problem. Suppose we have a coin that either is honest or is a coin that has been weighted so that when tossed its probability of coming up heads is .6. We wish to test whether the coin is honest or is the weighted coin by tossing it three times and observing the number of heads that is obtained. Our sample here is the triple of numbers (x_1, x_2, x_3), where $x_i = 1$ or 0 corresponding to whether a head or a tail was obtained on the ith toss. The sample space for this experiment consists of the eight points shown in Figure 1.

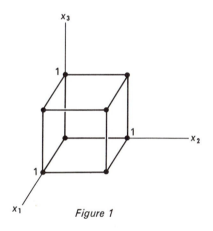

Figure 1

We may treat this as a problem of testing the hypothesis $H_0 \colon \theta = .5$ against $H_1 \colon \theta = .6$, where X is a Bernoulli random variable with parameter θ and from which a random sample of size three has been taken. Since more heads are likely to be obtained under H_1 than under H_0 and since $x_1 + x_2 + x_3$ represents the total number of heads that was obtained in the three tosses, let us choose the region A_1 to consist of those sample space points where $x_1 + x_2 + x_3$ is large. For example, we might choose the region A_1 to consist of the single point whose coordinates are $(1, 1, 1)$ and A_0 to consist of the seven remaining points. Or we might choose A_1 to consist of the four points $(1, 1, 1)$, $(1, 1, 0)$, $(1, 0, 1)$, and $(0, 1, 1)$ with A_0 consisting of the remaining four points.

If a_0 and a_1 denote the decisions to accept H_0 and H_1, respectively, our decision function $d = d(x_1, x_2, x_3)$ must map each of the eight sample points onto either a_0 or a_1. For A_1 chosen to be the point $(1, 1, 1)$, with A_0 consisting of the remaining seven points, our decision function would be determined by the values

$$d(0, 0, 0) = d(1, 0, 0) = d(0, 1, 0) = d(0, 0, 1) = d(1, 1, 0)$$
$$= d(1, 0, 1) = d(0, 1, 1) = a_0 \quad \text{and} \quad d(1, 1, 1) = a_1.$$

For testing problems it is usually simpler to look at the problem geometrically as one of deciding how to separate the sample space into two parts, rather than how to formally write out the decision function.

Our intuition would undoubtedly produce a good test for this simple illustrative problem, but there is no assurance that it would do so in a more complicated problem; therefore, we need to develop a theory based on the general principles described in Chapter 1. In this development we will begin by assuming that we are given the density function $f(x \mid \theta)$ of some random variable and that we wish to test the hypothesis $H_0: \theta = \theta_0$ against $H_1: \theta = \theta_1$ by means of a random sample $X_1, \ldots, X_n$. Let A_0 and A_1 denote the acceptance regions of sample space for the hypotheses H_0 and H_1, respectively, and let X now represent the vector random variable $X = (X_1, \ldots, X_n)$. In Chapter 1 the loss function for testing problems was chosen to be the function defined by

$$\mathcal{L}(\theta, d) = \begin{cases} 0, & \text{if the correct decision is made} \\ 1, & \text{if an incorrect decision is made.} \end{cases}$$

As a result, it follows that the two possible values of the risk function are given by

$$\mathcal{R}(\theta_0, d) = E\mathcal{L}(\theta_0, d(X)) = 0 \cdot P(X \in A_0 \mid \theta_0) + 1 \cdot P(X \in A_1 \mid \theta_0)$$

and

$$\mathcal{R}(\theta_1, d) = E\mathcal{L}(\theta_1, d(X)) = 1 \cdot P(X \in A_0 \mid \theta_1) + 0 \cdot P(X \in A_1 \mid \theta_1).$$

Hence,

(1) $$\mathcal{R}(\theta_0, d) = P(X \in A_1 \mid \theta_0)$$

and

$$\mathcal{R}(\theta_1, d) = P(X \in A_0 \mid \theta_1).$$

Both of these probabilities are probabilities of making an incorrect decision. In the first case H_0 is true, but if X falls in A_1 we decide incorrectly in favor of H_1. In the second case H_1 is true, but if X falls in A_0 we decide incorrectly in favor of H_0. The first of these two kinds of incorrect decisions is called a *type I error* and the second is called a *type II error*. The probability given in (1) for making such an error is called the *size of the error*. The size of the type I error is usually denoted by the letter α and the size of the type II error by β. These symbols and definitions may be summarized in the following manner.

Definition 1

(2) $\alpha = $ *size of type I error* $= P(Accept \ H_1 \mid H_0)$

$\beta = $ *size of type II error* $= P(Accept \ H_0 \mid H_1).$

Now we would like to choose a decision function $d = d(x_1, \ldots, x_n)$ that will separate the sample space into two parts A_0 and A_1 so as to simultaneously minimize $\mathcal{R}(\theta_0, d)$ and $\mathcal{R}(\theta_1, d)$. It is clear from (1), however, that the value of $\mathcal{R}(\theta_0, d)$ can be made to decrease in value only by shifting sample points from A_1 to A_0, and this will tend to increase the value of $\mathcal{R}(\theta_1, d)$. Since $\mathcal{R}(\theta_0, d)$ (or $\mathcal{R}(\theta_1, d)$) can be made to have the value zero by choosing A_0 (respectively, A_1) to be the entire sample space, it follows that there is no partition of the sample space that will simultaneously minimize $\mathcal{R}(\theta_0, d)$ and $\mathcal{R}(\theta_1, d)$, and therefore there cannot exist a best decision function d without further restrictions or conventions.

In Chapter 1 it was suggested that the introduction of an additional principle, such as that of minimax, might be necessary to arrive at a basis for determining a best decision function. In the chapter on estimation, the difficulty was circumvented in most situations by restricting estimates to being unbiased. The difficulty here can be eliminated by a simple convention without the necessity of introducing additional restrictive principles. It consists in agreeing to compare only those decision functions (by means of their $\mathcal{R}(\theta_1, d)$ values) that possess the same value of $\mathcal{R}(\theta_0, d)$. In terms of the notation in (2), we compare decision functions that possess the same α value by means of their β values.

Since a decision function merely determines how the sample space is to be separated into the two disjoint parts A_0 and A_1, we can just as well discuss the problem geometrically by concerning ourselves with how to choose A_0 and A_1. The union of A_0 and A_1 is the entire sample space; therefore, it suffices to concentrate on choosing only one of them, say A_1. It is traditional in the theory of testing hypotheses to call the region A_1 the *critical region for testing H_0*, rather than the acceptance region for H_1. The probability of the sample point falling in the critical region when H_0 is true, which has been denoted by α, is then called the *size of the critical region*. This refers to its probability size or weight, and has nothing to do with its geometrical size. In terms of this traditional language, the basic problem of testing a hypothesis H_0 is to find a critical region of size α that will minimize β. If there exists a critical region of size α that minimizes β among all critical regions whose size does not exceed α, it is called a *best critical region of size α*. The value of α is often chosen in advance, based upon practical considerations, and only critical regions of this size or less are permitted in the competition. A test that is based on such a best critical region is called a *best test of size α*.

As a summary of the preceding material and for reference in later work, the more important of the preceding definitions are displayed below.

> **Definition 2** *The critical region of a test is that part of sample space that corresponds to the rejection of the hypothesis H_0.*
>
> *The size of a critical region, α, is the probability of the sample point falling in the critical region when H_0 is true.*

> **Definition 3** *A best critical region of size α is one that minimizes the probability, β, of accepting H_0 when H_1 is true among all critical regions whose size does not exceed α.*
>
> *A best test is a test that is based on a best critical region.*

For the purpose of illustrating the preceding definitions, let X be a discrete random variable whose density function depends upon a parameter θ, and assume that the hypothesis $H_0: \theta = \theta_0$ is to be tested against the hypothesis $H_1: \theta = \theta_1$ on the basis of a single observed value of X. Let the density values of X for $\theta = \theta_0$ and $\theta = \theta_1$ be those given in the following table.

x	0	1	2	3	4	5
$f(x \mid \theta_0)$	.02	.03	.05	.05	.35	.50
$f(x \mid \theta_1)$	.04	.05	.08	.12	.41	.30

If we choose $\alpha = .05$, the possible critical regions of this size are the regions where (a) $x = 0$ or $x = 1$, (b) $x = 2$, (c) $x = 3$. The value of β corresponding to each of those critical regions is (a) .91, (b) .92, (c) .88. Among those three critical regions the critical region $x = 3$ is therefore the best. Before we can claim that it is the best critical region of size $\alpha = .05$, it is necessary to make certain that there is no critical region of size less than .05 that possesses a value of $\beta < .88$. The nontrivial critical regions of size less than .05 are (d) $x = 0$ and (e) $x = 1$. Since the corresponding values of β are (d) .96 and (e) .95, it follows that the test based on choosing $x = 3$ as critical region is a best test of size .05.

3.1. Neyman-Pearson Lemma

Fortunately, a method for constructing best critical regions is available, via a theorem called the Neyman-Pearson Lemma and named after the two statisticians who introduced and proved it. In this theorem it is assumed that the basic random variable X possesses the probability density $f(x \mid \theta)$ and that a random sample of size n is to be taken. A statement of the theorem and its proof for continuous X follow.

Neyman-Pearson Lemma   If there exists a critical region C of size α and a nonnegative constant k such that

$$(3) \qquad \frac{\prod_{i=1}^{n} f(x_i \mid \theta_1)}{\prod_{i=1}^{n} f(x_i \mid \theta_0)} \geq k \qquad \text{for points in } C$$

and

$$\frac{\prod_{i=1}^{n} f(x_i \mid \theta_1)}{\prod_{i=1}^{n} f(x_i \mid \theta_0)} \leq k \qquad \text{for points not in } C,$$

then C is a best critical region of size α.

Proof. In order to simplify the notation we shall let $L(x)$ with the proper subscript denote the likelihood function under each of the two hypotheses, and we shall let dx represent $dx_1 \cdots dx_n$ in multiple integration over sample space. Thus,

$$L_0(x) = \prod_{i=1}^{n} f(x_i \mid \theta_0) \quad \text{and} \quad L_1(x) = \prod_{i=1}^{n} f(x_i \mid \theta_1).$$

Now let C^* denote any other critical region of size $\leq \alpha$. The two critical regions C and C^* may be represented by the sets of points labeled C and C^* in Figure 2. Their intersection is denoted by e and their non-intersecting parts by a and b, respectively.

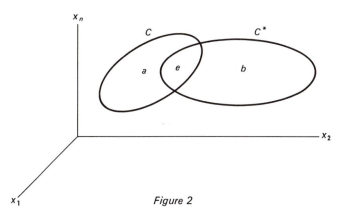

Figure 2

Since C and C^* are critical regions of sizes α and $\leq \alpha$, respectively, it follows from (1) and (2) that

(4)
$$\int_C \cdots \int \prod_{i=1}^{n} f(x_i \mid \theta_0) \, dx_1 \cdots dx_n = \int_C L_0(x) \, dx = \alpha$$

and

$$\int_{C^*} \cdots \int \prod_{i=1}^{n} f(x_i \mid \theta_0) \, dx_1 \cdots dx_n = \int_{C^*} L_0(x) \, dx \leq \alpha.$$

Hence,

(5)
$$\int_C L_0(x) \, dx \geq \int_{C^*} L_0(x) \, dx.$$

Writing $C = a + e$ and $C^* = b + e$, we may cancel the integral over e from both sides of (5) to reduce it to

(6)
$$\int_a L_0(x) \, dx \geq \int_b L_0(x) \, dx.$$

Let β and β^* denote the sizes of the type II error for the critical regions C and C^*, respectively. Since the size of a type II error is the probability that the sample point will fall outside the critical region when H_1 is true, which in turn is equal to 1 minus the probability that it will fall inside the critical region when H_1 is true, we may write

$$\beta = 1 - \int_C L_1(x) \, dx$$

and

$$\beta^* = 1 - \int_{C^*} L_1(x) \, dx.$$

Hence,

$$\beta^* - \beta = \int_C L_1(x) \, dx - \int_{C^*} L_1(x) \, dx.$$

Canceling the integral over the common part of C and C^* will give

(7)
$$\beta^* - \beta = \int_a L_1(x) \, dx - \int_b L_1(x) \, dx.$$

From the definition of C given in (3), it follows that $L_1(x) \geq kL_0(x)$ for all points in C, and hence for all points in a, and therefore that

$$\int_a L_1(x) \, dx \geq k \int_a L_0(x) \, dx.$$

Similarly, since b lies outside C, every point of b satisfies the second inequality of (3), namely $L_1(x) \leq kL_0(x)$; consequently

$$\int_b L_1(x) \, dx \leq k \int_b L_0(x) \, dx.$$

Applying these two results to (7) will yield the inequality

(8)
$$\beta^* - \beta \geq k \int_a L_0(x) \, dx - k \int_b L_0(x) \, dx.$$

But from (6) the right side must be nonnegative; therefore we arrive at the conclusion that

$$\beta^* \geq \beta.$$

Since β^* is the size of the type II error for any critical region, other than C, of size less than or equal to α, this proves that C is a best critical region of size α. ∎

There is a converse theorem which, when the notion of a best test is formulated more generally, essentially states that a best critical region must be of the type described by the Neyman-Pearson Lemma.

As an illustration of how this theorem enables us to find a best critical region, consider the problem of testing the hypothesis $H_0: \mu = \mu_0$ against

$H_1: \mu = \mu_1 > \mu_0$, where μ is the mean of a normal density with known variance σ^2. Here

$$L = \prod_{i=1}^{n} \frac{\exp\left[-\frac{1}{2}\left(\frac{x_i - \mu}{\sigma}\right)^2\right]}{\sqrt{2\pi}\sigma} = \frac{\exp\left[-\frac{1}{2\sigma^2}\Sigma(x_i - \mu)^2\right]}{(2\pi\sigma^2)^{n/2}}.$$

Hence,

(9)
$$\frac{L_1}{L_0} = \frac{\exp\left[-\frac{1}{2\sigma^2}\Sigma(x_i - \mu_1)^2\right]}{\exp\left[-\frac{1}{2\sigma^2}\Sigma(x_i - \mu_0)^2\right]}$$

$$= \exp\left[-\frac{1}{2\sigma^2}\left[\Sigma(x_i - \mu_1)^2 - \Sigma(x_i - \mu_0)^2\right]\right]$$

$$= \exp\left[-\frac{1}{2\sigma^2}\left[2(\mu_0 - \mu_1)\Sigma x_i + n(\mu_1^2 - \mu_0^2)\right]\right]$$

$$= \exp\left[\frac{\mu_1 - \mu_0}{\sigma^2}\Sigma x_i + \frac{n(\mu_0^2 - \mu_1^2)}{2\sigma^2}\right].$$

From (3) it now follows that the critical region C will be determined by the inequality

$$\exp\left[\frac{\mu_1 - \mu_0}{\sigma^2}\Sigma x_i + \frac{n(\mu_0^2 - \mu_1^2)}{2\sigma^2}\right] \geq k,$$

where k is some nonnegative constant. Taking logarithms will yield the equivalent inequality

$$\frac{\mu_1 - \mu_0}{\sigma^2}\Sigma x_i \geq \log k + \frac{n(\mu_1^2 - \mu_0^2)}{2\sigma^2}.$$

It was assumed under H_1 that $\mu_1 > \mu_0$; therefore the factor in front of Σx_i on the left is positive, and so our inequality reduces to

$$\Sigma x_i \geq \frac{\sigma^2 \log k}{\mu_1 - \mu_0} + \frac{n(\mu_1 + \mu_0)}{2}.$$

Since k may be chosen to be any nonnegative number, as k ranges over values from 0 to ∞ the right side of this inequality will assume values from $-\infty$ to $+\infty$; therefore this inequality is equivalent to the inequality

(10)
$$\Sigma x_i \geq a,$$

where a may be chosen to be any real number. The equation $\Sigma x_i = a$ is that of a plane in n dimensional sample space. In Figure 3 the part of this plane with positive coordinates is sketched.

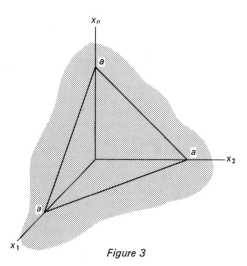

Figure 3

The critical region C is therefore that part of sample space which lies above this plane. As a assumes increasingly large numerical values, the region C includes increasingly less of the sample space, and as a goes from $-\infty$ to ∞, C shifts from including all of the sample space to none of it. Thus, it is clear that by a proper choice of a, which is equivalent to a proper choice of k, we can choose C to be a region of any desired probability size α, other than 0 or 1.

Now the value of α that is selected affects only the value of the constant a and not the shape of the critical region C; therefore a test based on this type of critical region must be a best test of its size, whatever the size. Furthermore, since $\sum x_i \geq a$ is equivalent to $\bar{x} \geq a/n$, the best test here is equivalent to one that is based on the random variable $\bar{X}$ and which chooses as critical region the interval $\bar{X} \geq b$, where b is a constant selected to satisfy $P(\bar{X} \geq b \mid H_0) = \alpha$.

The striking feature of this test is its simplicity, in that its critical region can be made to depend only upon the single random variable $\bar{X}$ whose properties are well known, rather than upon the n dimensional random variable $(X_1, \ldots, X_n)$.

As a numerical illustration, consider the problem that was discussed in the introduction to this chapter, namely that of testing whether a shipment of nylon fishing line is 8 pound or 10 pound test line. Suppose experience has shown that the breaking strength of such line may be treated as a normal variable with $\sigma = 2$, and suppose that a random sample of size 16 yielded $\bar{X} = 9$. The problem then is to test the hypothesis $H_0: \mu = 8$ against $H_1: \mu = 10$ by means of this information. We shall choose $\alpha = .05$. Since the best test here is based on the critical region

$\overline{X} \geq b$ where b is chosen to satisfy $P(\overline{X} \geq b \mid \mu = 8) = .05$, it is necessary to find the correct value of b. Now when $\mu = 8$, $\overline{X}$ is a normal variable with mean 8 and with standard deviation $\dfrac{\sigma}{\sqrt{n}} = \dfrac{2}{\sqrt{16}} = .5$; it therefore follows that $\tau = \dfrac{\overline{X} - 8}{.5}$ is a standard normal variable when H_0 is true. Hence

$$P(\overline{X} \geq b \mid \mu = 8) = P\left(\frac{\overline{X} - 8}{.5} \geq \frac{b - 8}{.5}\right)$$

$$= P\left(\tau \geq \frac{b - 8}{.5}\right).$$

From Table IV in the appendix it will be found that $P(\tau \geq 1.64) = .05$; consequently b must be chosen to satisfy the equation $\dfrac{b - 8}{.5} = 1.64$, which is equivalent to choosing $b = 8.82$. Our critical region of size .05 therefore consists of those sample points for which $\overline{X} \geq 8.82$. The sample value $\overline{X} = 9$ falls in this critical region; hence H_0 is rejected in favor of H_1. This problem will be discussed more thoroughly in the next section.

As a second illustration of how best tests are constructed, consider the problem of testing the hypothesis $H_0 : \sigma = 10$ against $H_1 : \sigma = 8$ for a normal variable with mean zero, if $\alpha = .025$ and if a sample of size 20 yielded $\sum_{i=1}^{20} x_i^2 = 1600$.

For the purpose of obtaining a more general test, we shall solve the problem of testing $H_0 : \sigma = \sigma_0$ against $H_1 : \sigma = \sigma_1 < \sigma_0$ for a normal variable with known mean μ based on a random sample of size n. The best critical region

$$\frac{L_1}{L_0} = \frac{\sigma_0^n \exp\left[-\dfrac{\sum (x_i - \mu)^2}{2\sigma_1^2}\right]}{\sigma_1^n \exp\left[-\dfrac{\sum (x_i - \mu)^2}{2\sigma_0^2}\right]} \geq k$$

is equivalent to the region

$$\exp\left[\frac{1}{2}\left(\frac{1}{\sigma_0^2} - \frac{1}{\sigma_1^2}\right) \sum (x_i - \mu)^2\right] \geq k\left(\frac{\sigma_1}{\sigma_0}\right)^n.$$

Since $\sigma_1 < \sigma_0$, this reduces to

$$\sum (x_i - \mu)^2 \leq b,$$

where b is some constant. When H_0 is true, we know from Section 2.7 that the variable $\dfrac{\sum (X_i - \mu)^2}{\sigma_0^2}$ possesses a chi-square distribution with n

degrees of freedom; hence the best test for this problem consists of choosing as critical region those sample points for which

$$\frac{\sum (x_i - \mu)^2}{\sigma_0^2} \leq \chi_0^2,$$

where χ_0^2 is a value such that $P(\chi^2 \leq \chi_0^2) = \alpha$.

For the preceding numerical problem, it will be found from Table V in the appendix that $\chi_0^2 = 9.59$ for $\alpha = .025$ and 20 degrees of freedom. Since $\mu = 0$ for our problem, the critical region for this test is given by the inequality

$$\frac{\sum x_i^2}{100} \leq 9.59.$$

Our sample value $\sum x_i^2 = 1600$ does not satisfy this inequality; therefore H_0 is accepted here.

As an illustration for a discrete variable, consider the problem of testing $H_0: \mu = \mu_0$ against $H_1: \mu = \mu_1 < \mu_0$, where μ is the mean of a Poisson variable. Here

$$\frac{L_1}{L_0} = \frac{\prod\limits_{i=1}^{n} e^{-\mu_1} \frac{\mu_1^{x_i}}{x_i!}}{\prod\limits_{i=1}^{n} e^{-\mu_0} \frac{\mu_0^{x_i}}{x_i!}} = e^{n(\mu_0 - \mu_1)} \left(\frac{\mu_1}{\mu_0}\right)^{\Sigma x_i}.$$

The critical region C given by (3) then becomes the region determined by

$$e^{n(\mu_0 - \mu_1)} \left(\frac{\mu_1}{\mu_0}\right)^{\Sigma x_i} \geq k,$$

which is equivalent to

$$\sum x_i \cdot \log \frac{\mu_1}{\mu_0} \geq \log k + n(\mu_1 - \mu_0).$$

Since $\frac{\mu_1}{\mu_0} < 1$, this is equivalent to

$$\sum x_i \leq \frac{\log k}{\log \frac{\mu_1}{\mu_0}} + \frac{n(\mu_1 - \mu_0)}{\log \frac{\mu_1}{\mu_0}}.$$

The same reasoning as for the preceding illustrations will show that the critical region $\sum x_i \leq b$, where b is a constant, is a best critical region of its size for testing H_0 against H_1.

It was shown in Volume I that the sum of a set of independent Poisson variables is a Poisson variable with mean equal to the sum of the individual means; therefore the variable $Z = \sum_{i=1}^{n} X_i$ is a Poisson variable with mean $n\mu_0$ when H_0 is true. The preceding test is therefore equivalent to the test based on the Poisson variable Z with mean $n\mu_0$, which chooses as critical region those values of Z that satisfy an inequality of the type $Z \leq b$.

We may run into difficulties when attempting to apply the preceding test. Suppose that we had chosen $\alpha = .10$ and that $n\mu_0$ had the value 3. Under H_0 the probability density of the Poisson variable Z would then be given by

$$f(z) = \frac{e^{-3} \cdot 3^z}{z!}.$$

Calculations will yield the approximate values $f(0) = .05$, $f(1) = .15$, $f(2) = .22, \dots$. It is not possible to find a constant b to satisfy the requirement $P(Z \leq b) = .10$ because this probability will be equal to .05 for $b < 1$ and will be equal to .20 for $1 \leq b < 2$. Thus there is no value of b that will yield a probability of exactly .10. This difficulty, which usually arises for discrete variable problems, can be circumvented by introducing a *randomization device*. Here it would consist in agreeing to always reject H_0 if $Z = 0$, which event will occur with probability .05; but to reject H_0 only one third of the time if $Z = 1$, which event will occur with probability .15. Thus, if we should obtain the value $Z = 1$, we would employ a game of chance for which the probability of success is one third and reject H_0 if, and only if, the game of chance produced a success. A test of this type is called a *randomized test* because it introduces a random device to determine the critical region. In general, if the critical region is of the type $Z \leq b$, we would start with the sample point farthest to the left and keep adding sample points to the critical region as we moved to the right, until we arrived at a point such that the total probability mass accumulated was less than or equal to α and such that this probability would be greater than α if we added one more point. This additional point would be assigned to the critical region only part of the time. The probability that it would be assigned is determined so as to produce a total probability of α for the critical region. In practice, such randomization techniques are seldom used. It is customary to use a critical region that does not require randomization but whose value is close to the desired value of α. The preceding discussion is of theoretical interest, however, as it shows that predetermined values of α can be attained for discrete variables.

As a second numerical illustration for a discrete variable, consider the problem of testing $H_0: p = .4$ against $H_1: p = .3$ for a Bernoulli variable, if $\alpha = .05$ and if 50 trials of the experiment produced 16 successes. The likelihood function here is given by

$$L = \prod_{i=1}^{50} p^{x_i}(1-p)^{1-x_i} = p^S(1-p)^{50-S},$$

where $S = \sum x_i$ denotes the total number of successes obtained in the 50 trials. Then

$$\frac{L_1}{L_0} = \frac{(.3)^S(.7)^{50-S}}{(.4)^S(.6)^{50-S}} = \left(\frac{7}{6}\right)^{50}\left(\frac{9}{14}\right)^S.$$

A best critical region is therefore of the form $\left(\frac{9}{14}\right)^S \geq b$, which is equivalent to $S \leq a$ for the proper a. As in the case of the Poisson variable discussed in the preceding paragraph, it is unlikely that a critical region of exact size $\alpha = .05$ can be obtained without a randomization device. S is the observed value of a binomial variable, which will be denoted by X. Therefore to find such a best critical region, it would be necessary to calculate the values of the binomial density

$$f(x \mid .4) = \binom{50}{x}(.4)^x(.6)^{50-x},$$

beginning with $x = 0$ and continuing to $x = a - 1$ until $\sum_{x=0}^{a-1} f(x \mid .4) \leq .05$ but such that adding $f(a \mid .4)$ would produce a value larger than .05. Then it would be possible to introduce randomization at the value $x = a$ to obtain a critical region of size .05. Now the binomial variable X is known to possess a distribution that can be approximated very well by means of a normal distribution with $\mu = np$ and $\sigma^2 = np(1 - p)$ when $p = .4$ and $n = 50$; therefore we shall solve this problem by means of the normal approximation and thereby eliminate the necessity for the lengthy computations needed to solve the problem by the preceding exact method. The variable $\tau = \dfrac{(X - np)}{\sqrt{np(1 - p)}} = \dfrac{(X - 20)}{\sqrt{12}}$ may be treated as an approximate standard normal variable when $p = .4$. Hence, the constant a must be chosen to satisfy

$$P(X \leq a) = P\left(\frac{X - 20}{\sqrt{12}} \leq \frac{a - 20}{\sqrt{12}}\right) = .05.$$

From Table IV in the appendix $P(\tau \leq -1.64) = .05$; therefore a must satisfy the equation $\dfrac{a - 20}{\sqrt{12}} = -1.64$. The solution of this equation is $a = 14.3$; consequently the critical region consists of those values of X satisfying $X \leq 14.3$. Since the sample value $S = 16$ does not lie in the critical region, H_0 is accepted in preference to H_1. The error arising in using the normal approximation for a binomial distribution is very small for a value of n as large as 50 when $p = .4$; therefore the critical region $X \leq 14.3$ will correspond closely to the more exact critical region obtained by using discrete methods and randomization.

3.1.1. Calculation of β.   Thus far our illustrations have been concerned only with how to construct a best critical region. Now let us consider the numerical problem of how to evaluate β. For the example concerned with testing a normal mean, assume once more that $\mu_0 = 8$, $\mu_1 = 10$, $\sigma = 2$, $n = 16$, and $\alpha = .05$. The critical region for that problem was found to be $\bar{X} \geq 8.82$; therefore

$$\beta = P(\bar{X} < 8.82 \mid \mu = 10).$$

Here $\bar{X}$ is a normal variable with mean 10 and standard deviation .5; hence $\tau = \dfrac{\bar{X} - 10}{.5}$ is a standard normal variable. From Table IV in the appendix we then obtain

$$\beta = P\left(\frac{\bar{X} - 10}{.5} < \frac{8.82 - 10}{.5}\right)$$

$$= P(\tau < -2.36) = .009.$$

The geometrical meanings of α and β for this problem are displayed in Figure 4 where the two normal $\bar{X}$ distributions under H_0 and H_1 are sketched.

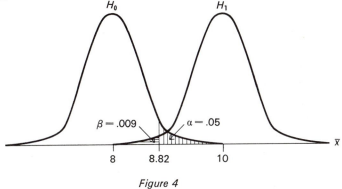

Figure 4

The calculation of β for a discrete variable such as the Poisson variable of our second earlier illustration is not as easy as for a continuous variable, particularly if we use a randomized test. For that earlier problem we chose $\alpha = .10$ and assumed that the variable $Z = \sum_{i=1}^{n} X_i$ possessed the mean 3 under H_0. For purposes of illustration, assume that the mean is 2 under H_1. The critical region of that test consisted of the point $Z = 0$ and the randomized point $Z = 1$, which was assigned to the critical region with probability $1/3$. It is simpler here to calculate the value of $1 - \beta$ than to calculate the value of β directly. Thus,

$$1 - \beta = P(Z \in \text{critical region} \mid H_1).$$

Since the Poisson variable Z has mean 2 under H_1, its distribution is given by the density

$$f(z) = \frac{e^{-2}2^z}{z!}.$$

The probability that Z will fall in the preceding randomized critical region is therefore given by

$$f(0) + (1/3)f(1) = e^{-2} + (1/3)e^{-2} \cdot 2 = .23.$$

Consequently, $\beta = .77$ for this problem.

In the first of the two preceding problems we chose $\alpha = .05$ and found that $\beta = .009$, whereas in the second problem we chose $\alpha = .10$ and found that $\beta = .77$. Now in any real-life problem the choice of α should be governed by the importance of the two types of error. If, for example, an experimenter requires strong protection against making a type I error, the value of α should be chosen small enough to satisfy him. However, if he is equally concerned about making a type II error, then the critical region should be chosen so as to produce approximately equal values of α and β. In our first illustration, $\beta = .009$ is so small compared to $\alpha = .05$ that it would seem advisable to decrease the value of α if comparable size errors are desired. In the second illustration it would be necessary to increase α considerably to make the magnitude of β comparable to that of α. Since we are not in a position to evaluate the relative importance of the two types of errors in problems, we shall usually choose $\alpha = .05$. This is the value of α that is chosen by many experimenters because of tradition. Therefore if the value of α is not specified in a problem, it should be understood that $\alpha = .05$.

In view of the difficulties that can arise when the size of the sample is chosen in advance of the calculations of α and β, it is advisable to determine how large n must be in order to yield desired values of α and β. As an illustration of how this is done, consider the problem of determining the value of n that will yield $\alpha = .01$ and $\beta = .03$ for the problem related to Figure 4. If the critical region is to be given by $\overline{X} \geq c$, then c and n must satisfy the two equations

$$.01 = P(\overline{X} > c \mid \mu = 8)$$

$$.03 = P(\overline{X} < c \mid \mu = 10).$$

These equations are equivalent to the equations

$$.01 = P\left(\frac{(X - 8)\sqrt{n}}{2} > \frac{(c - 8)\sqrt{n}}{2}\right) = P\left(\tau > \frac{(c - 8)\sqrt{n}}{2}\right)$$

$$.03 = P\left(\frac{(X - 10)\sqrt{n}}{2} < \frac{(c - 10)\sqrt{n}}{2}\right) = P\left(\tau < \frac{(c - 10)\sqrt{n}}{2}\right),$$

where τ is a standard normal variable. From Table IV in the appendix these equations will be satisfied if

$$\frac{(c - 8)\sqrt{n}}{2} = 2.33 \quad \text{and} \quad \frac{(c - 10)\sqrt{n}}{2} = -.188.$$

Eliminating c will yield the equation

$$8 + 2.33 \frac{2}{\sqrt{n}} = 10 - 1.88 \frac{2}{\sqrt{n}}.$$

Solving for n will give $n = 17.7$; hence a sample of size 18 should suffice to yield the desired values of α and β. The sample already taken is therefore nearly large enough to produce these values of α and β.

Although the Neyman-Pearson Lemma was derived on the assumption that $f(x \mid \theta)$ is the probability density of a single random variable X with a single parameter θ, there is nothing in the derivation that precludes X and θ from being vectors; therefore our theory of best tests applies to problems in any dimension.

3.2. Composite hypotheses

If a hypothesis specifies the value of the parameter θ it is called a *simple hypothesis*, otherwise it is called a *composite hypothesis*. . The hypotheses H_0 and H_1 of the preceding illustrations were simple hypotheses. When θ is a vector parameter, a simple hypothesis must specify the values of all its components. Thus, for a normal density it would be necessary to specify the value of both μ and σ. In our first illustration the value of σ was assumed known; therefore it was a single parameter problem. If σ had not been known, the hypothesis $H_0: \mu = \mu_0$ would have been a composite hypothesis.

A composite hypothesis that occurs frequently because of its practical importance is a hypothesis of the type $H: \mu > \mu_0$. For example, suppose a certain ingredient is added to gasoline with the hopes of increasing mileage. If a set of experiments is conducted to determine whether the ingredient does increase mileage, these experiments can be analyzed by treating the problem as one of testing the hypothesis

(11) $\qquad H_0: \mu = \mu_0 \qquad$ against $\qquad H_1: \mu > \mu_0.$

Here μ_0 represents the average mileage for the car used in the experiment when it uses gasoline without the added ingredient. This composite alternative has the distinct advantage of not requiring us to know what the true average mileage is when using the modified gasoline. Long experience often yields a value of μ_0, but this is not true for the alternative value μ_1 when it corresponds to an experimental mean.

A striking feature of the best test obtained in our first illustration of Section 3.1 is that it does not depend upon the numerical value of μ_1, provided that $\mu_1 > \mu_0$. This is because the nature of the critical region was determined by an inequality of the type $(\mu_1 - \mu_0) \sum x_i \geq c$, and this inequality reduces to an inequality of the type $\sum x_i \geq a$ regardless of the numerical value of μ_1 as long as $\mu_1 > \mu_0$. The particular value of a needed to produce a critical region of size α depends upon the value of μ_0, but not on the value of μ_1. In view of this, it follows that our test is a best test for the problem given in (11). Thus, although the Neyman-Pearson Lemma was designed to produce a best test for testing a simple hypothesis against a simple alternative, it produced a best test for testing a simple hypothesis against a composite alternative hypothesis. The best tests that we obtained for the other illustrations also possess this broader coverage, since in each case the nature of the critical region did not depend upon the particular value of the parameter specified under H_1.

Although the preceding formulation of a testing problem suffices for many practical problems, there are situations in which it would be preferable to test the hypothesis

(12) $H_0 : \mu \leq \mu_0$ against $H_1 : \mu > \mu_0$.

Here we have a composite hypothesis H_0 as well as a composite alternative H_1. It is rather satisfying to know that the best tests which we constructed by means of the Neyman-Pearson Lemma for the preceding illustrations are also best for composite hypotheses of the type (12) for those problems.

A critical region of size α for this composite H_0 requires that the probability of a sample point falling in the critical region shall be less than or equal to α for all values of μ satisfying $\mu \leq \mu_0$. This property of a best test for testing $H_0 : \mu = \mu_0$ against $H_1 : \mu = \mu_1 > \mu_0$ also being best for the composite hypotheses listed in (12), is known to hold for many of the familiar probability distributions such as the normal, binomial, Poisson, and chi-square distributions. The applicability of the Neyman-Pearson Lemma is therefore much broader than one might suppose from the statement of the lemma.

All of the preceding tests possess the property of being one-sided with respect to the alternative hypothesis H_1. That is, the parameter under H_1 is permitted to assume values only to the left, or only to the right, of the values permitted under H_0. Unless we have reason to believe that a parameter must have a smaller, or larger, value under H_1 than the value hypothesized under H_0, we should prefer a two-sided alternative. For example, in the illustration of testing whether the addition of an ingredient to gasoline had increased its mileage, we assumed that there had been either no change or an improvement; however, if we had no valid basis for expecting an improvement, we might prefer to test the hypothesis of no change against the alternative of some change, either good or bad. Our test would then be of the form

(13) $H_0 : \mu = \mu_0$ against $H_1 : \mu \neq \mu_0$.

From our earlier illustration on testing a normal mean, it will be observed that the best test exclusively used the right tail of the $\overline{X}$ distribution as critical region when $\mu_1 > \mu_0$ and, by symmetry, the left tail if $\mu_1 < \mu_0$. Symmetry considerations would require that the critical region consist of equal tails of the $\overline{X}$ distribution when testing (13), and therefore we should not expect to obtain a best test when our alternative value μ_1 can be either smaller or larger than μ_0. Best tests for the testing problem (13) do not exist for the familiar probability distributions, unless further restrictions are placed on the nature of the tests.

As a numerical illustration of a type of problem that occurs frequently and to which (13) is often applied, consider the problem of testing the hypothesis $H_0: p_1 = p_2$ against $H_1: p_1 \neq p_2$, where p_1 and p_2 are the probabilities of success in two different experiments. Suppose that 100 trials of the first experiment yielded 40 successes, and 150 trials of the second experiment yielded 50 successes, and that we have chosen $\alpha = .10$. Since the values of n here are sufficiently large to insure good normal approximations to the corresponding binomial variables, we may assume that the two sample success ratios $\hat{p}_1 = X_1/n_1$ and $\hat{p}_2 = X_2/n_2$ are independently normally distributed with means p_1 and p_2 and with variances $p_1(1 - p_1)/n_1$ and $p_2(1 - p_2)/n_2$. Hence, when H_0 is true the variable $\hat{p}_1 - \hat{p}_2$ may be assumed to be normally distributed with mean 0 and with variance $\dfrac{p_1(1 - p_1)}{n_1} + \dfrac{p_2(1 - p_2)}{n_2}$. As a result, the variable

$$Z = \frac{\hat{p}_1 - \hat{p}_2}{\sqrt{\dfrac{p_1(1 - p_1)}{n_1} + \dfrac{p_2(1 - p_2)}{n_2}}}$$

may be treated as a standard normal variable. Since the alternative hypothesis is symmetric with respect to p_1 and p_2, we shall choose as critical region the two .05 tails of the distribution of Z. Thus our critical region will be given by

$$\left| \frac{\hat{p}_1 - \hat{p}_2}{\sqrt{\dfrac{p_1(1 - p_1)}{n_1} + \dfrac{p_2(1 - p_2)}{n_2}}} \right| > 1.64.$$

Even this approximate test cannot be carried out because it requires a knowledge of p_1 and p_2; however, since approximations have already been introduced, we shall introduce one more in order to obtain a workable test. When H_0 is true $p_1 = p_2$; therefore, if p denotes this common but unknown value, it may be approximated by the combined sample estimate which is $\hat{p} = \dfrac{X_1 + X_2}{n_1 + n_2}$. If this value is substituted for p_1 and p_2 in the preceding test, our critical region will be given by

$$\left| \frac{\hat{p}_1 - \hat{p}_2}{\sqrt{\hat{p}(1 - \hat{p}) \left[\dfrac{1}{n_1} + \dfrac{1}{n_2} \right]}} \right| > 1.64.$$

Calculations based on our sample values will give the estimate $\hat{p} = \frac{90}{250} = .36$. Since $\hat{p}_1 = .40$ and $\hat{p}_2 = .33$,

$$\frac{\hat{p}_1 - \hat{p}_2}{\sqrt{\hat{p}(1 - \hat{p}) \left[\dfrac{1}{n_1} + \dfrac{1}{n_2} \right]}} = \frac{.07}{\sqrt{(.36)(.64) \left[\dfrac{1}{100} + \dfrac{1}{150} \right]}} = 1.13.$$

This value does not lie in the critical region; consequently H_0 is accepted in preference to H_1.

This test is only an approximate test because it not only employs the normal approximation for a binomial distribution but also replaces the unknown parameter p by its sample estimate. The error arising from using the normal approximation is very small for sample sizes as large as these; however, the error that may occur from replacing p by its sample approximation can be a more serious matter, although for samples as large as this it also will be small.

It is unrealistic to believe that the probabilities of success in two different experiments are identical in value; therefore in accepting a hypothesis such as H_0 the experimenter does not imply that he believes H_0 is true. He is willing to accept, however, that the difference in those probabilities is so small that it may be treated as being equal to zero, at least as far as his experiments are concerned. It would be more realistic in a problem such as this to test the hypothesis

$$H_0: |p_1 - p_2| \le \varepsilon \qquad \text{against} \qquad H_1: |p_1 - p_2| > \varepsilon,$$

where ε is chosen as the smallest difference that is worth distinguishing; however, it is more difficult mathematically to construct a test for this formulation of the problem than it is for our standard version.

3.2.1. The power function.   In Section 3.1.1 the method of calculating the value of β was illustrated on two problems that involved testing a simple hypothesis against a simple alternative. If the alternative hypothesis is composite, such as in (13), the size of the type II error will depend upon what particular value of θ in the alternative domain of θ values is being considered. Thus, the problem of determining the size of the type II error for a test involving a composite alternative becomes the problem of determining the function $\beta(\theta)$ that gives the size of the type II error for every possible value of θ that can occur under H_1.

Since the calculation of α requires the integration or summation of $f(x \mid \theta_0)$ over the critical region, it is usually more convenient to calculate $1 - \beta(\theta)$ than $\beta(\theta)$ because it also requires integration or summation over the critical region. The function $1 - \beta(\theta)$, which is called the power function of the test, will be denoted by $P(\theta)$. For convenience of reference the following formal definition of the power function is given.

Definition 4 *The power function $P(\theta)$ of a test is the function of the parameter that gives the probability that the sample point will fall in the critical region of the test, and hence the probability that the hypothesis H_0 will be rejected, when θ is the true value of the parameter.*

By studying the power function it is possible to determine how good a test is in detecting various alternative values of θ when they represent the true situation. If we were testing a simple hypothesis against a simple alternative, we would be interested only in the two values $P(\theta_0) = \alpha$ and $P(\theta_1) = 1 - \beta$. In terms of this notation, the test given by the Neyman-Pearson Lemma is a test that possesses maximum power at θ_1 among all tests for which $P(\theta_0) \leq \alpha$. For a problem such as (13), we should try to find a test that possesses maximum power for all alternative values of θ among tests for which $P(\theta_0) \leq \alpha$. As was pointed out earlier, we cannot expect to find such a test for problems like (13) but can do so for some of the one-sided problems of the type (11). A test is called a *uniformly most powerful test* if among all tests satisfying $P(\theta_0) \leq \alpha$ its power function is a maximum for all values of θ in the domain determined by H_1. As we observed earlier, the test based on $\bar{X} \geq c$ was a best test for testing a normal mean for problem (11); therefore that test is a uniformly most powerful test for that problem. The test based on using equal tails of the $\bar{X}$ distribution for problem (13) is, however, not such a test.

For the purpose of studying the properties of the test that uses equal tails of the $\bar{X}$ distribution for problem (13), let us calculate the power function of the test for the data used in the illustration associated with Figure 4. Since $\alpha = .05$ and we are to use equal tails of the $\bar{X}$ distribution for our critical region, we must select the constant c to satisfy

$$P(\bar{X} > c \mid \mu = 8) = .025.$$

Then, just as in the earlier problem, c must satisfy

$$P\left(\frac{\bar{X} - 8}{.5} > \frac{c - 8}{.5}\right) = .025,$$

or

$$P\left(\tau > \frac{c - 8}{.5}\right) = .025,$$

where τ is a standard normal variable when H_0 is true. From Table IV in the appendix it will be found that c must satisfy $\dfrac{c-8}{.5} = 1.96$, so that $c = 8 + .98 = 8.98$. By symmetry, the critical region therefore consists of the two intervals $\bar{X} < 7.02$ and $\bar{X} > 8.98$. Then, from Definition 4, it follows that

$$P(\theta) = P(\bar{X} \in \text{critical region} \mid \mu = \theta)$$

$$= \int_{-\infty}^{7.02} \frac{\exp\left[-\dfrac{1}{2}\left(\dfrac{\bar{x}-\theta}{.5}\right)^2\right]}{\sqrt{2\pi}(.5)} \, d\bar{x} + \int_{8.98}^{\infty} \frac{\exp\left[-\dfrac{1}{2}\left(\dfrac{\bar{x}-\theta}{.5}\right)^2\right]}{\sqrt{2\pi}(.5)} \, d\bar{x}.$$

Letting $\tau = \dfrac{\bar{x}-\theta}{.5}$ gives

$$P(\theta) = \int_{-\infty}^{(7.02-\theta)/.5} \frac{e^{-\tau^2/2}}{\sqrt{2\pi}} \, d\tau + \int_{(8.98-\theta)/.5}^{\infty} \frac{e^{-\tau^2/2}}{\sqrt{2\pi}} \, d\tau.$$

By assigning values to θ and using Table IV in the appendix, a sufficient number of values of $P(\theta)$ can be obtained to produce a smooth graph of $P(\theta)$. The graph of $P(\theta)$, which is the symmetrical one shown in Figure 5, was obtained in this manner.

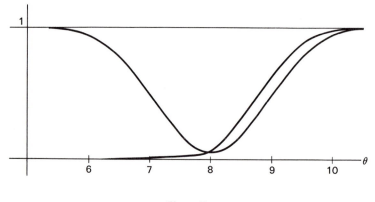

Figure 5

For the purpose of comparing this two-sided test with the uniformly most powerful one-sided test for problem (11), which is the test that uses the interval $\bar{X} > 8.82$ as its critical region, the graph of the power function for that test is also shown in Figure 5. Its graph lies above that for the two-sided test for values of $\theta > 8$, which are the only alternative values permitted in (11). It will be observed from these graphs that the relative advantage of the best one-sided test is most pronounced for values of $\theta > 8$ which are not too far removed from $\theta = 8$.

3.3. Sequential tests

Thus far we have been concerned only with how to design a good test for testing some hypothesis against an alternative hypothesis on the basis of a random sample of size n. If we are given a small sample and have selected our α value in advance to be small, say $\alpha = .05$, we may discover that our β value for the best test that can be constructed is embarrassingly large. This implies, for example, that if we are testing to see whether a new product is equal, or else superior, to an old product in quality, our probability will be very large of overlooking the superior product. This large value of β can be decreased by increasing the size of α. However, even after the value of α has been adjusted to produce a more compatible value of β, we may find that both values are much too large to satisfy us. The only way out of such a dilemma would appear to be to take an additional sample. To guard against such contingencies, it would seem advisable to choose values of both α and β in advance and then determine how large a sample will be needed to attain those error rates. The experimenter who proceeds in this intelligent manner will often discover to his dismay that a very large value of n is needed to guarantee the values of α and β that he selected. Although there seems to be little that he can do other than increase his values of α and β, there are in fact techniques available for helping him to decrease the size of n somewhat while keeping his α and β, provided that he is willing to gamble a bit more.

Suppose we are testing whether a shipment of some product is up to quality specifications or has fallen below a certain tolerance level of quality. Instead of taking a fixed sample, say 100, which had been determined as being necessary to guarantee our choices of α and β, we could take samples one at a time and test them until we had accumulated enough positive information to make a decision. If the shipment was one of inferior quality we might be able to discover that very quickly, and therefore it would be unnecessary to test all 100 items. Similarly, if the quality of the shipment were high, we might be able to discover that quickly also. Thus, by capitalizing on the possibilities of very bad or very good shipments, we should be able to decrease the sample size needed to arrive at a decision and still be assured that our values of α and β are realized. A sampling scheme of this type in which the samples are taken one at a time is called *sequential sampling*, and a test based on this type of sampling is called a *sequential test*. The new element that has been added here is that of postponing a positive decision until sufficient information has been accumulated to produce a verdict with guaranteed reliability. Thus, at each stage of the sampling we can decide to accept H_0, or accept H_1, or take an additional sample.

It has been shown mathematically that, for given values of α and β, the size of the sample needed to arrive at a decision of accepting $H_0: \theta = \theta_0$ or $H_1: \theta = \theta_1$ is on the average considerably smaller than the fixed sample size, n_0, needed to do so for the standard problems of statistics. For small values of α and β, say in the neighborhood of .05, the mean sample size is often about $(1/2)n_0$. Since the value of n is uncertain in any given sequential problem, there is the possibility that a very large n will be needed to arrive at a positive decision. Therefore the experimenter must be prepared to take this additional gamble. The probability that his value of n will exceed the fixed size sample needed to produce the same α and β values can be calculated, however, and it usually is very small for standard problems.

For the purpose of describing a sequential test, we shall assume that X possesses a density $f(x \mid \theta)$ and that we wish to test $H_0: \theta = \theta_0$ against $H_1: \theta = \theta_1$. Since we are testing a simple hypothesis against a simple alternative, the Neyman-Pearson Lemma would suggest using the likelihood ratio

$$\frac{L_{1n}}{L_{0n}} = \frac{\prod_{i=1}^{n} f(x_i \mid \theta_1)}{\prod_{i=1}^{n} f(x_i \mid \theta_0)}$$

as a basis for making a decision. If the sample size n is fixed, this lemma places all sample points for which the likelihood ratio is larger than a certain constant k in the acceptance region for H_1, and all points for which it is smaller than k in the acceptance region for H_0. A sequential test can be constructed by extending this method slightly to include a region for deciding to continue sampling. It would appear that we should choose as the region for continuing to sample those points for which the likelihood ratio is neither much larger nor much smaller than k. Thus if we wish to design a sequential test that capitalizes on the known optimal properties of the likelihood ratio for constructing a fixed size sample test, we should choose two positive constants c_0 and c_1 and agree to continue sampling as long as

(14) $$c_0 < \frac{L_{1n}}{L_{0n}} < c_1.$$

However, at the first n for which $\dfrac{L_{1n}}{L_{0n}} \leq c_0$ we should stop and accept H_0,

and at the first n for which $\dfrac{L_{1n}}{L_{0n}} \geq c_1$ we should stop and accept H_1. A

particular sequential test of this type, which will now be described in some detail, has been shown to possess certain optimal properties in the class of all sequential tests.

It can be shown that if c_0 and c_1 are chosen properly, the preceding sequential type test can be made to have prescribed values of α and β. Precise formulas for determining c_0 and c_1 as functions of α and β are not available; however, excellent approximations to the exact values are given by choosing

$$(15) \qquad c_0 = \frac{\beta}{1 - \alpha} \quad \text{and} \quad c_1 = \frac{1 - \beta}{\alpha}.$$

Some plausible arguments will be given in the next section to justify these approximations and to indicate the extent of their accuracy. The name given to this test and the technique for carrying it out when these approximations are used will be expressed in the following form.

Sequential Probability Ratio Test To test the hypothesis $H_0: \theta = \theta_0$ *against* $H_1: \theta = \theta_1$, *calculate the likelihood ratio* $\dfrac{L_{1n}}{L_{0n}}$ *and proceed as follows:*

$$(16) \qquad \text{(i) if } \frac{L_{1n}}{L_{0n}} \leq \frac{\beta}{1 - \alpha}, \quad \text{accept } H_0;$$

$$\text{(ii) if } \frac{L_{1n}}{L_{0n}} \geq \frac{1 - \beta}{\alpha}, \quad \text{accept } H_1;$$

$$\text{(iii) if } \frac{\beta}{1 - \alpha} < \frac{L_{1n}}{L_{0n}} < \frac{1 - \beta}{\alpha}, \quad \text{take an additional sample.}$$

A remarkable feature of this test is that it does not require any knowledge of the distribution of a statistic in order to carry it out. In two of the illustrations of the Neyman-Pearson Lemma that we considered earlier, it was necessary to know the distribution of $\overline{X}$ for a normal distribution and the distribution of $Z = \sum X_i$ for a Poisson distribution before the critical region of a desired size could be determined. This distribution was also necessary if the value of β was to be calculated. In the sequential test α and β are selected in advance at any desired levels, and then the test is carried out by merely calculating the likelihood ratio for successively larger samples and observing which of the inequalities in (16) is satisfied.

As an example of how the sequential test proceeds, consider the same problem that was used to illustrate the Neyman-Pearson Lemma, namely that of testing $H_0: \mu = \mu_0$ against $H_1: \mu = \mu_1$ where μ is the mean of a normal variable with known variance. The likelihood ratio was computed earlier in (9) to be

$$\frac{L_{1n}}{L_{0n}} = \exp\left[\frac{\mu_1 - \mu_0}{\sigma^2} \sum x_i + \frac{n(\mu_0^2 - \mu_1^2)}{2\sigma^2} \right].$$

For this problem the test becomes simpler if we replace the inequalities of (16) by the equivalent inequalities that are obtained by taking logarithms.

Then inequality (iii) of (16) reduces to

$$\log \frac{\beta}{1-\alpha} < \frac{\mu_1 - \mu_0}{\sigma^2} \sum x_i + \frac{n(\mu_0^2 - \mu_1^2)}{2\sigma^2} < \log \frac{1-\beta}{\alpha}.$$

Since $\mu_1 > \mu_0$, this is equivalent to

$$(17) \qquad \frac{\sigma^2}{\mu_1 - \mu_0} \log \frac{\beta}{1-\alpha} + \frac{n(\mu_1 + \mu_0)}{2} < \sum x_i$$

$$< \frac{\sigma^2}{\mu_1 - \mu_0} \log \frac{1-\beta}{\alpha} + \frac{n(\mu_1 + \mu_0)}{2}.$$

If we had assumed that $\mu_1 < \mu_0$, these inequalities would have been reversed.

As a numerical example, suppose we have chosen $\alpha = .05$ and $\beta = .10$, and that we are testing $\mu_0 = 10$ against $\mu_1 = 10.5$ with $\sigma = 1$. Then (17) becomes

$$-4.50 + 10.25n < \sum x_i < 5.78 + 10.25n,$$

and the sequential test proceeds as follows:

(i) if $\displaystyle\sum_{i=1}^{n} x_i \leq -4.50 + 10.25n$, accept $\mu = 10$;

(18)   (ii) if $\displaystyle\sum_{i=1}^{n} x_i \geq 5.78 + 10.25n$, accept $\mu = 10.5$;

(iii) if neither inequality is satisfied, take another observation.

An experiment was performed by taking successive samples of a normal variable with mean 10 and variance 1 until a positive decision was reached. The decision to accept $\mu = 10$, which is the correct decision here, occurred at the fifteenth sample. The fifteen values of X_i obtained in the sampling were the following: 10.91, 8.88, 10.35, 9.84, 10.42, 10.05, 9.94, 11.15, 9.92, 10.23, 10.53, 8.61, 9.70, 8.83, 8.96. The values of $\sum x_i$ together with the values of the decision boundaries given in (18) are shown in Table 1.

Calculations will show that a value of $n = 34$ would be needed to guarantee values of $\alpha = .05$ and $\beta = .10$ for the optimal fixed sample size test for this problem. Calculations based on formula (23), which will be presented in the next section, show that the expected value of n for this sequential problem is 16. Thus there is a large reduction in the sample size required for this problem if it is treated sequentially.

The preceding theory was described for a single random variable X possessing a density $f(x \mid \theta)$ that depended on a single parameter θ; however, the theory applies equally well to vector variables and vector parameters. The only restriction is that the hypotheses H_0 and H_1 must be simple hypotheses.

Table 1

n	1	2	3	4	5	6	7
b_0	5.75	16.00	26.25	36.50	46.75	57.00	67.25
$\sum x_i$	10.91	19.79	30.14	39.98	50.40	60.45	70.39
b_1	16.03	26.28	36.53	46.78	57.03	67.28	77.53

n	8	9	10	11	12	13	14	15
b_0	77.50	87.77	98.00	108.25	118.50	128.75	139.00	149.25
$\sum x_i$	81.54	91.46	101.69	112.22	120.83	130.53	139.36	148.32
b_1	87.78	98.03	108.28	118.53	128.78	139.03	149.28	159.53

Although practical problems of testing seldom fit into a test in which both H_0 and H_1 are simple hypotheses, it is often possible to modify a problem so that such a test will prove satisfactory. For example, suppose that we wish to test whether the proportion of voters who favor a presidential candidate is the same as it was a month ago, or whether it has increased during the past month. Thus, we would like to test $H_0: p = p_0$ against $H_1: p > p_0$, where p_0 is the earlier proportion. Since a very small increase over p_0 would not be very meaningful, we might well agree on a proportion $p_1 > p_0$ that is sufficiently larger than p_0 to be worthy of notice and then test $H_0: p = p_0$ against $H_1: p = p_1$. In this form we could apply the sequential probability ratio test with prescribed values of α and β and arrive at a decision that would be meaningful in a practical sense.

3.3.1. Approximations for c_0 and c_1.   The purpose of this section is to give a heuristic justification for the values of c_0 and c_1 employed in the sequential probability ratio test. Toward this objective, for each value of $n = 1, 2, 3, \ldots$, let the corresponding n dimensional sample space be divided into the three regions R_{0n}, R_{1n}, and R_{an} corresponding to the three possible decisions of accepting H_0, accepting H_1, or taking another sample, respectively. If the sample point falls in R_{0n} or R_{1n} for some value of n, it must have fallen in R_{an} for all smaller values of n, otherwise H_0 or H_1 would have been accepted earlier.

Now there is no fixed value of n such that a decision to accept H_0 or H_1 must occur before then; therefore from a theoretical point of view it is necessary to consider infinite dimensional sample spaces here. Since they are considerably more sophisticated than the finite sample spaces that we have employed thus far, we shall simplify our development by assuming that finite space methods may be applied here. We shall also assume that the probability is one of eventually accepting H_0 or H_1.

If the sample point is denoted by X regardless of the dimension of the sample space and if we attempt to calculate the quantity $1 - \beta$, we would write

$$1 - \beta = P(\text{accept } H_1 \mid H_1)$$

(19)
$$= \sum_{n=1}^{\infty} P(X \in R_{1n} \mid H_1)$$

$$= \sum_{n=1}^{\infty} \int_{R_{1n}} \cdots \int L_{1n} \, dx_1 \cdots dx_n.$$

Since the sequential test based on c_0 and c_1 before the approximations (15) were introduced yields a point in R_{1n} only if that point satisfies the inequality $\dfrac{L_{1n}}{L_{0n}} \geq c_1$, this inequality may be applied to the integrand in (19) to give

$$1 - \beta \geq \sum_{n=1}^{\infty} \int_{R_{1n}} \cdots \int c_1 L_{0n} \, dx_1 \cdots dx_n$$

$$= c_1 \sum_{n=1}^{\infty} \int_{R_{1n}} \cdots \int L_{0n} \, dx_1 \cdots dx_n$$

$$= c_1 \sum_{n=1}^{\infty} P(X \in R_{1n} \mid H_0)$$

$$= c_1 \, P(\text{accept } H_1 \mid H_0)$$

$$= c_1 \alpha.$$

Hence, if $\alpha > 0$, the number c_1 must satisfy the inequality

(20)
$$c_1 \leq \frac{1 - \beta}{\alpha}.$$

A similar type of calculation applied to the evaluation of β will yield the inequality

(21)
$$c_0 \geq \frac{\beta}{1 - \alpha},$$

provided that $\alpha < 1$.

Now assume that α and β are both small, say less than .10 in value each, and consider the graphs in the α, β plane of the two lines whose equations are

(22) $c_1 \alpha = 1 - \beta$ and $c_0(1 - \alpha) = \beta.$

The graphs of these lines for a typical pair of values of c_0 and c_1 are shown in Figure 6.

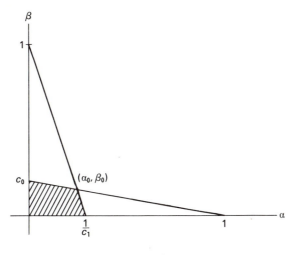

Figure 6

Inequalities (20) and (21) are equivalent to the inequalities $c_1 \alpha \le 1 - \beta$ and $c_0(1 - \alpha) \ge \beta$, which will be satisfied by only those points lying in the shaded area of Figure 6. This means that if values of c_0 and c_1 are selected, the sizes of the two types of error cannot exceed the values corresponding to points in the shaded area. If we now choose the numbers α_0 and β_0 which we would like to have for our error sizes, and use formula (15) to determine c_0 and c_1 for our test, we know that the actual values of α and β cannot exceed those for points in the shaded area of Figure 6 where $c_0 = \dfrac{\beta_0}{1 - \alpha_0}$ and $c_1 = \dfrac{1 - \beta_0}{\alpha_0}$. Since the point of intersection of lines (22) satisfies equations (22), and since α_0 and β_0 also satisfy those equations, this point of intersection has coordinates (α_0, β_0). Thus, it follows from Figure 6 that the actual values of α and β cannot be much larger than the approximate values α_0 and β_0 given by the point of intersection. Furthermore, if the value of α exceeds the value given by this point, then the corresponding value of β must be less than that given by this point. Similarly, if the value of β is larger than β_0, the corresponding value of α must be smaller than α_0. For sequential tests in which α_0 and β_0 are chosen quite small and for which the mean sample size required to reach a positive decision is not too small, the difference in the actual α and β values and the values α_0 and β_0 selected in advance to determine c_0 and c_1 by formula (15) are negligible.

Because the sequential test was introduced as a means of decreasing the sample size needed for testing, it may be appropriate here to write down the formula for $E(n)$, but omitting its lengthy derivation. The designer of a sampling experiment can use the formula to decide whether the saving on mean sample size is sufficient to compensate for the inconvenience in taking samples one at a time and for the uncertainty that a decision may not be reached before a sample considerably larger than $E(n)$ is taken. This formula gives only an approximation to the value of $E(n)$. The value of $E(n)$ depends upon the true value of θ in the problem; therefore this dependence is shown as a subscript.

$$(23) \qquad E_\theta(n) \doteq \frac{P(H_1 \mid \theta) \log c_2 + [1 - P(H_1 \mid \theta)] \log c_1}{E_\theta[\log f(X \mid \theta_1) - \log f(X \mid \theta_0)]}.$$

Here $P(H_1 \mid \theta)$ is the probability of accepting H_1 when θ is the true value of the parameter; hence, for example, $P(H_1 \mid \theta_0) = \alpha$ and $P(H_1 \mid \theta_1) = 1 - \beta$. This formula is the one that was used in the numerical illustration of the preceding section to arrive at the value $E(n) = 16$ for that problem.

The sequential probability ratio test (16) has been shown mathematically to possess the remarkable optimal property that among all sequential tests with the same α and β values, it minimizes the two values $E_{\theta_0}(n)$ and $E_{\theta_1}(n)$. This is truly remarkable in that the test simultaneously minimizes these two vital quantities.

3.4.   Likelihood ratio tests

The methods of testing hypotheses that have been presented thus far enable us to solve problems of testing a simple hypothesis against a simple alternative. For some special problems these methods are also applicable to testing composite hypotheses; however, for the more general problem we do not yet have a systematic method for testing such hypotheses. What is needed then to round out our methods is a test for composite hypotheses.

Since a simple hypothesis may be treated as a special case of a composite hypothesis, the most general problem that can arise is one of testing a composite hypothesis against a composite alternative. For the purpose of formulating such a problem, let the set of possible values of the parameter θ, which may be a vector, be denoted by Θ. This set is called the *parameter space*, just as the set of possible sample points defines the sample space. Any hypothesis about θ merely restricts its values in some manner, which means that θ can assume values only in some subset of Θ. A pair of composite hypotheses, H_0 and H_1, can therefore be expressed in the form

$$(24) \qquad\qquad H_0 : \theta \in \Theta_0 \qquad \text{and} \qquad H_1 : \theta \in \Theta_1,$$

where Θ_0 and Θ_1 are two disjoint subsets of Θ.

As an illustration, consider the problem of testing the hypothesis $H_0 : \mu = \mu_0$ against $H_1 : \mu \neq \mu_0$, where μ is the mean of a normal variable with unknown variance σ^2. Here the vector parameter $\theta = (\mu, \sigma)$ is conveniently represented geometrically as a point in two dimensions. If we choose a horizontal axis to represent μ values and a vertical axis to represent σ values, the parameter space Θ will be represented by the upper half-plane because σ cannot assume negative values. Here Θ_0 will be represented by the vertical half-line whose equation is $\mu = \mu_0$, and Θ_1 will be represented by $\Theta - \Theta_0$, that is, by the upper half-plane with the half-line $\mu = \mu_0$ deleted. These geometrical representations are shown in Figure 7.

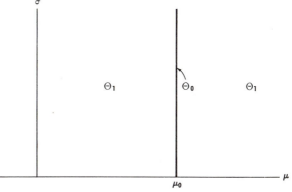

Figure 7

We have seen that the likelihood ratio principle leads to best tests when testing simple hypotheses, both for a fixed size sample and for sequential sampling, and that it even leads to best tests for some special types of composite hypotheses; therefore we should try to incorporate it into a test for general composite hypotheses. If we denote the likelihood functions corresponding to H_0 and H_1 in (24) by $L(x \mid \Theta_0)$ and $L(x \mid \Theta_1)$, we should attempt to use the likelihood ratio $\dfrac{L(x \mid \Theta_1)}{L(x \mid \Theta_0)}$ as a basis for constructing a test; however, this ratio is a function of the unspecified parameters in the two hypotheses as well as of the random sample values. It is only when H_0 and H_1 are simple hypotheses that the ratio is independent of unknown parameter values. Therefore, we must replace unknown parameters by their sample estimates, or by other known values, if this ratio is to serve as a basis for constructing a test. In view of the optimal properties of maximum likelihood estimates, an obvious solution would seem to be to replace all unknown parameters by their maximum likelihood estimates. For general hypotheses of the type (24), however, the likelihood function, even if it is a continuous function of its arguments, does not always attain a maximum in the domain of θ values determined by the hypothesis. The maximum may occur on the boundary, and the boundary may not belong to the domain. If these maxima do occur inside their respective domains there is no difficulty, and we could then use the quantity

$$\frac{L(x \mid \hat{\theta}_1)}{L(x \mid \hat{\theta}_0)}$$

as a basis for constructing the test, where $\hat{\theta}_0$ and $\hat{\theta}_1$ indicate that the unknown parameters under H_0 and H_1 have been replaced by their corresponding maximum likelihood estimates. By anology with earlier tests, we would choose as our critical region those sample points where

$$\frac{L(x \mid \hat{\theta}_1)}{L(x \mid \hat{\theta}_0)} \geq k,$$

where k is a suitably chosen constant.

The problems that will be studied in this book, which include a large share of the standard testing problems that arise in statistical applications, are of the type in which $\Theta_1 = \Theta - \Theta_0$ and in which the dimension of Θ_0 is lower than that of Θ. This was true, for example, for the illustration concerning a normal mean displayed in Figure 7. For that problem it is clear that the maximum likelihood estimates $\hat{\mu}$ and $\hat{\sigma}$ when μ and σ are restricted to lie in Θ_1 will be the same as the maximum likelihood estimates $\bar{x}$ and $[\sum (x_i - \bar{x})^2/n]^{1/2}$ when μ and σ are free to vary throughout Θ. The only time this will not be true is when $\bar{x}$ has the value μ_0, because then the maximizing point will lie in Θ_0 rather than in Θ_1; however, for a continuous random variable such as $\overline{X}$, the probability of that occurring is zero. Thus, we could just as well replace $\hat{\theta}_1$ by $\hat{\theta}$ in the likelihood ratio for this problem, where $\hat{\theta}$ denotes the unrestricted maximum likelihood estimate of θ.

Since we will be dealing exclusively with problems of the preceding type for which the maximum likelihood estimates $\hat{\theta}_0$, $\hat{\theta}_1$, and $\hat{\theta}$ exist and for which the probability that $\hat{\theta}_1 = \hat{\theta}$ is one, we may replace $\hat{\theta}_1$ by $\hat{\theta}$ in our likelihood ratio. With this replacement, the value of the likelihood ratio must be at least one. It is more convenient to work with a ratio whose value lies between 0 and 1 rather than between 1 and ∞; therefore it is traditional in likelihood ratio tests to invert the preceding ratio and base the test on the inverted ratio. Thus, our likelihood ratio test will choose as its critical region those values of the random variable λ satisfying $0 \leq \lambda \leq \lambda_0$ where

$$\lambda = \frac{L(X \mid \hat{\theta}_0)}{L(X \mid \hat{\theta})}$$

and where λ_0 is a suitably chosen constant.

Even though λ is a random variable that depends only upon the random vector $X = (X_1, \ldots, X_n)$, there is no assurance that its probability distribution will not depend upon the unknown parameters occurring in $f(x \mid \theta)$ when H_0 is true. If its distribution is independent of such parameters, we can find a numerical value λ_0 such that $P(\lambda \leq \lambda_0) = \alpha$, and our test will be complete. Fortunately, in many important applications λ turns out to be a familiar random variable whose distribution is known and for which the value of λ_0 can be found. In case the distribution of λ does depend upon unknown parameters, the probability $P(\lambda \leq \lambda_0)$ will depend upon those parameters. The value of λ_0 must then be chosen so that this probability will not exceed α regardless of the values of the unknown parameters under H_0.

The preceding discussion should help motivate the following likelihood ratio test for composite hypotheses. It should be understood, however, that this test is not capable of treating the most general composite hypotheses as expressed in (24). The desired test for our restricted class of problems can be summarized as follows.

Likelihood Ratio Test *To test the composite hypothesis $H_0: \theta \in \Theta_0$ against $H_1: \theta \in \Theta - \Theta_0$, use the statistic λ given by*

(25)
$$\lambda = \frac{L(X \mid \hat{\theta}_0)}{L(X \mid \hat{\theta})}$$

where $\hat{\theta}_0$ and $\hat{\theta}$ represent the maximum likelihood estimators of the unspecified parameters in Θ_0 and Θ, respectively; and reject H_0 if, and only if, the sample value of λ satisfies the inequality $\lambda \leq \lambda_0$ where λ_0 is a value such that $P(\lambda \leq \lambda_0 \mid H_0) \leq \alpha$ for all θ under H_0.

Although the distribution of λ can often be found, many problems arise in which the derivation of the distribution is exceedingly difficult. Fortunately, if n is large and certain conditions are satisfied, there exists an approximation to the distribution of λ that is satisfactory for most large sample applications of the test. This property may be expressed as follows.

Property 1 *Under the proper regularity conditions on $f(x \mid \theta)$, the random variable $-2 \log_e \lambda$ will possess an asymptotic chi-square distribution with degrees of freedom equal to the difference between the number of independent parameters in Θ and Θ_0.*

In the normal distribution illustration of Figure 7, for example, the parameter space Θ was two-dimensional and Θ_0 was one-dimensional; hence $-2 \log \lambda$ for that problem would possess an asymptotic chi-square distribution with one degree of freedom. As another illustration, suppose the random variable X possesses the density $f(x \mid \theta_1, \ldots, \theta_k)$, and that H_0 is the hypothesis $H_0: \theta_1 = \cdots = \theta_k$. Then if f satisfies the required regularity conditions, $-2 \log \lambda$ will possess an asymptotic chi-square distribution with $k - 1$ degrees of freedom because there will be only one independent parameter unspecified under H_0.

As an illustration of how the asymptotic distribution performs on a familiar problem, let us apply it to testing $H_0: \mu = \mu_0$ against $H_1: \mu \neq \mu_0$ for a normal variable with known σ based on a random sample of size n. The standard symmetric test for this problem chooses as critical region the two tails of the $\overline{X}$ distribution. If $\alpha = .05$, these are given by $|\overline{X} - \mu_0| > 1.96(\sigma/\sqrt{n})$.

Here the likelihood functions are

$$L(x \mid \Theta) = \frac{\exp\left[-\dfrac{1}{2\sigma^2}\sum(x_i - \mu)^2\right]}{(2\pi)^{n/2}\sigma^n}$$

and

$$L(x \mid \Theta_0) = \frac{\exp\left[-\dfrac{1}{2\sigma^2}\sum(x_i - \mu_0)^2\right]}{(2\pi)^{n/2}\sigma^n}.$$

Replacing μ by its maximum likelihood estimate $\bar{x}$ and performing some algebraic simplifications on the likelihood ratio will produce the value

$$\lambda = \exp\left[-\frac{n(\bar{x} - \mu_0)^2}{2\sigma^2}\right].$$

Hence,

$$-2 \log \lambda = \frac{n(\bar{x} - \mu_0)^2}{\sigma^2}.$$

Accordingly, $n(\overline{X} - \mu_0)^2/\sigma^2$ should possess an asymptotic chi-square distribution with one degree of freedom. However, since $\sqrt{n}(\overline{X} - \mu_0)/\sigma$ is a standard normal variable when H_0 is true, we know that $n(\overline{X} - \mu_0)^2/\sigma^2$ possesses an exact chi-square distribution with one degree of freedom. Thus, the approximation here for large n happens to be exact. Now the critical region for the likelihood ratio test is given by $\lambda \leq \lambda_0$, which is equivalent to $-2 \log \lambda \geq -2 \log \lambda_0 = \chi_0^2$, where $P(\chi^2 > \chi_0^2) = .05$. Since $-2 \log \lambda = n(\overline{X} - \mu_0)^2/\sigma^2$, our critical region can be expressed in the form $|\overline{X} - \mu_0| \geq \dfrac{\sigma}{\sqrt{n}}\sqrt{\chi_0^2}$. Thus our earlier symmetric two-sided test is the same as this test.

Next, let us apply the asymptotic theory to a problem for which we do not have the exact solution. Consider the problem of testing $H_0: \mu = 5$ against $H_1: \mu \neq 5$ where μ is the mean of a Poisson density, if a sample of size 12 produced the following sample values: 7, 3, 4, 8, 5, 6, 5, 12, 9, 7, 10, 4. Here

$$L(x \mid \Theta) = \prod_{i=1}^{12} \frac{e^{-\mu}\mu^{x_i}}{x_i!} = \frac{e^{-12\mu}\mu^{\sum x_i}}{\Pi x_i!}$$

and

$$L(x \mid \Theta_0) = \prod_{i=1}^{12} \frac{e^{-5} 5^{x_i}}{x_i!} = \frac{e^{-60} 5^{\Sigma x_i}}{\Pi x_i!}.$$

Since $\hat{\mu} = \bar{x}$, λ will assume the form

$$\lambda = \frac{e^{-60} 5^{12\bar{x}}}{e^{-12\bar{x}} (\bar{x})^{12\bar{x}}} = \left[\frac{e^{-5} 5^{\bar{x}}}{e^{-\bar{x}} (\bar{x})^{\bar{x}}} \right]^{12}.$$

Hence,

$$-2 \log \lambda = -24[-5 + \bar{x} + \bar{x} \log 5 - \bar{x} \log \bar{x}].$$

Calculations will show that $\bar{x} = 20/3$, and hence that $-2 \log \lambda = 6.0$. Since there is one unspecified parameter in Θ and none in Θ_0, the random variable $-2 \log \lambda$ will have an asymptotic chi-square distribution with one degree of freedom. From Table V in the appendix the .05 critical point is 3.84; therefore, since $6.0 > 3.84$, the hypothesis H_0 is rejected here.

This problem could have been solved by using the Poisson variable $Z = \sum_{i=1}^{12} X_i$ whose mean is 60 when H_0 is true and by choosing as critical region the two .025 tails of the Poisson distribution. An approximation to that method of solution using the Central Limit Theorem can be obtained by treating Z as an approximate normal variable with mean 60 and variance 60. Thus, we treat $\tau = (Z - \mu_0)/\sqrt{\mu_0}$ as a standard normal variable and use the two .025 tails as our critical region. Here $\tau = (80 - 60)/\sqrt{60} = 2.58$; therefore, since $2.58 > 1.96$, the hypothesis H_0 is rejected.

For the purpose of comparing these two approximate tests, we shall calculate $P(-2 \log \lambda > 6.0)$ and $P(|\tau| > 2.58)$ because they give the probabilities of obtaining a result further out in the respective critical regions than our sample result. These respective probabilities are .015 and .010; therefore the two approximate solutions differ very little here.

The preceding general likelihood ratio test was derived on somewhat intuitive grounds based on the optimal properties of likelihood ratio tests for simple hypotheses; nevertheless it does possess some optimal properties of its own. It can be shown that likelihood ratio tests are asymptotically optimal under the proper restrictions. The definition of asymptotic optimality employed in that demonstration is quite sophisticated and therefore will not be discussed here. It has also been found that the likelihood ratio test often gives rise to best tests for some composite hypotheses for finite n when the class of tests is restricted further in a natural manner. Thus, the likelihood ratio function appears once more to be the backbone of good testing procedures.

3.4.1.   Student's t distribution.   As a first application of the likelihood ratio test given by (25), consider the problem that was discussed earlier of testing the hypothesis $H_0: \mu = \mu_0$ against $H_1: \mu \neq \mu_0$ where μ is the mean of a normal variable with unknown variance σ^2. Here

$$(26) \qquad L(x \mid \Theta) = \frac{\exp\left[-\dfrac{1}{2\sigma^2} \sum (x_i - \mu)^2\right]}{(2\pi)^{n/2}\sigma^n}$$

and

$$(27) \qquad L(x \mid \Theta_0) = \frac{\exp\left[-\dfrac{1}{2\sigma^2} \sum (x_i - \mu_0)^2\right]}{(2\pi)^{n/2}\sigma^n}.$$

In the preceding chapter it was shown that the joint maximum likelihood estimates of μ and σ are given by $\hat{\mu} = \bar{x}$ and $\hat{\sigma}^2 = \dfrac{\sum (x_i - \bar{x})^2}{n}$; hence these are the estimates needed to obtain $L(x \mid \hat{\theta})$. If these estimates are substituted into (26) we will obtain

$$L(x \mid \hat{\theta}) = \frac{e^{-n/2}}{(2\pi)^{n/2}} \left[\frac{\sum (x_i - \bar{x})^2}{n}\right]^{-n/2}.$$

The maximum likelihood estimate of σ in (27) is readily seen to be given by $\hat{\sigma}^2 = \dfrac{\sum (x_i - \mu_0)^2}{n}$; hence this will give

$$L(x \mid \hat{\theta}_0) = \frac{e^{-n/2}}{(2\pi)^{n/2}} \left[\frac{\sum (x_i - \mu_0)^2}{n}\right]^{-n/2}.$$

The likelihood ratio (25) for a set of random sample values then assumes the form

$$\lambda = \left[\frac{\sum (x_i - \mu_0)^2}{\sum (x_i - \bar{x})^2}\right]^{-n/2}.$$

The critical region $\lambda \leq \lambda_0$ is equivalent to the region $\lambda^{-2/n} \geq c$, where $c = \lambda_0^{-2/n}$. In view of these equivalence relations, the critical region of this test reduces to the region where

$$(28) \qquad \frac{\sum (x_i - \mu_0)^2}{\sum (x_i - \bar{x})^2} \geq c.$$

Some algebraic manipulations on the left side of (28) will reduce it to a more recognizable form. Thus,

$$\sum (x_i - \mu_0)^2 = \sum [(x_i - \bar{x}) + (\bar{x} - \mu_0)]^2 = \sum (x_i - \bar{x})^2 + n(\bar{x} - \mu_0)^2.$$

When this is applied to (28), we obtain the inequality

$$(29) \qquad \frac{n(\bar{x} - \mu_0)^2}{\sum (x_i - \bar{x})^2} \geq c - 1.$$

This in turn is equivalent to the inequality

(30)
$$\frac{n(\bar{x} - \mu_0)^2(n - 1)}{\sum (x_i - \bar{x})^2} \geq (c - 1)(n - 1).$$

At this stage it is necessary to refer to a distribution, called Student's t distribution, that was derived in Volume I. There it was shown that the variable t can be expressed in the form

(31)
$$t = \frac{u\sqrt{v}}{v},$$

where u is a standard normal variable, v^2 is a chi-square variable with v degrees of freedom, and u and v^2 are independent random variables. Table VI in the appendix gives values of $P(t > t_0)$ for various values of v and t_0. This variable has a distribution very close to that of a standard normal variable, and for values of $v > 20$ the distributions are virtually indistinguishable.

For the purpose of applying this distribution to (30), it suffices to choose

$$u = \frac{(\bar{X} - \mu_0)\sqrt{n}}{\sigma}, \qquad v^2 = \frac{\sum (X_i - \bar{X})^2}{\sigma^2}, \qquad \text{and} \qquad v = n - 1.$$

Under H_0 the mean of $\bar{X}$ is μ_0, and its variance if σ^2/n; therefore the variable u here qualifies as a standard normal variable. Furthermore, it will be shown in Chapter 5 that $\dfrac{\sum (X_i - \bar{X})^2}{\sigma^2}$ possesses a chi-square distribution with $n - 1$ degrees of freedom and that these two variables are independent random variables. We may therefore apply (31) to obtain the Student t variable given by

(32)
$$t = \frac{\sqrt{n}(\bar{X} - \mu_0)\sqrt{n - 1}}{\sqrt{\sum(X_i - \bar{X})^2}}.$$

A comparison of this value with that in (30) shows that our likelihood ratio test is equivalent to a test which is based on t^2 of the Student t statistic given by (32), and which uses as critical region equal tails of the Student t distribution with $n - 1$ degrees of freedom. Since the critical region given by (30) is equivalent to the region $t^2 \geq (c - 1)(n - 1)$, it suffices to find a value t_0 such that $P(t > t_0) = \alpha/2$, and then the critical region will consist of the two tail regions where $|t| > t_0$.

As a numerical illustration, consider a modification of the problem that was solved in Section 3.1 to illustrate the use of the Neyman-Pearson Lemma. There the problem was one of testing $H_0: \mu = 8$ against $H_1: \mu = 10$ by means of a random sample of size 16 which yielded the value $\bar{X} = 9$.

It was assumed that X possesses a normal distribution with $\sigma = 2$. We shall modify the problem by assuming that we do not know the value of σ but that our sample yielded the estimate $s = 2$, and that we wish to test $H_0: \mu = 8$ against $H_1: \mu \neq 8$. Since $s^2 = \dfrac{\Sigma (x_i - \bar{x})^2}{n}$, application of (32) gives

$$t = \frac{(9 - 8)\sqrt{15}}{\sqrt{4}} = 1.94.$$

With $\alpha = .05$ and $v = n - 1 = 15$ degrees of freedom, it will be found in Table VI in the appendix that the critical value of t here is given by $t_0 = 2.13$. The critical region of this test therefore consists of those sample points for which $|t| > 2.13$. Since $t = 1.94$ does not lie in the critical region, H_0 is accepted.

It is interesting to compare this result with the earlier result where it was assumed that $\sigma = 2$. The basic variable there was the standard normal variable $\tau = \dfrac{(\bar{X} - \mu)\sqrt{n}}{\sigma}$, whereas here it is the Student t variable which can be written in the form $t = \dfrac{(\bar{X} - \mu)\sqrt{n}}{\tilde{\sigma}}$, where $\tilde{\sigma}^2 = \dfrac{\Sigma(X_i - \bar{X})^2}{(n - 1)}$ is the traditional unbiased estimator of σ^2. In that earlier problem the critical region of the test was $|\tau| > 1.96$, as compared to our critical region here of $|t| > 2.13$. For n as large as 16, the distributions of τ and t differ very little; therefore these two tests differ very little when testing the same set of hypotheses. The test based on the t distribution possesses the advantage that it does not require a knowledge of σ, which is usually unavailable in real-life problems; consequently it is likely to be the more useful test.

Another use for Formula (32) is that of finding a confidence interval for a normal mean when the value of σ is not available. As an illustration, consider the problem that was discussed in Section 2.7, where a sample of size 25 yielded the value $\bar{X} = 20$ and σ was assumed to be equal to 1. The method introduced in that section produced the 95 percent confidence interval given by (19.6, 20.4). We shall modify this problem by assuming that σ is unknown, but that the sample yielded the value $s = 1$ as the estimate for σ. Since $v = 24$ here, Table VI in the appendix and (32) show that

$$P(|(\bar{X} - \mu)\sqrt{24}| < 2.064) = .95.$$

This is equivalent to

$$P\left(\bar{X} - \frac{2.064}{\sqrt{24}} < \mu < \bar{X} + \frac{2.064}{\sqrt{24}}\right) = .95.$$

Since $\bar{X} = 20$, the desired 95 percent confidence interval is therefore (19.58, 20.42), which, it will be observed, is slightly longer than the earlier interval. This should come as no surprise because it should be possible to estimate μ with greater precision if σ is known.

As another important application of the likelihood ratio test, consider the problem of testing whether the means of two independent normal variables that possess the same variances are equal. Let X and Y denote these variables, and assume that samples of sizes n_1 and n_2, respectively, are taken of them. If μ_1 and μ_2 denote their means and σ^2 is their common but unknown variance, it will follow that

$$(33) \quad L(x, y \mid \Theta) = \frac{\exp\left[-\dfrac{1}{2\sigma^2}\left[\displaystyle\sum_1^{n_1}(x_i - \mu_1)^2 + \sum_1^{n_2}(y_i - \mu_2)^2\right]\right]}{(2\pi)^{(n_1+n_2)/2}\sigma^{n_1+n_2}}$$

and

$$(34) \quad L(x, y \mid \Theta_0) = \frac{\exp\left[-\dfrac{1}{2\sigma^2}\left[\displaystyle\sum_1^{n_1}(x_i - \mu)^2 + \sum_1^{n_2}(y_i - \mu)^2\right]\right]}{(2\pi)^{(n_1+n_2)/2}\sigma^{n_1+n_2}},$$

where μ denotes their common but unknown mean under H_0. Here we are testing $H_0\colon \mu_1 = \mu_2$ against $H_1\colon \mu_1 \neq \mu_2$.

Letting $n = n_1 + n_2$ and omitting obvious summation indices, the usual logarithmic differentiation technique for finding maximum likelihood estimates when applied to (33) will give the estimates

$$\hat{\mu}_1 = \bar{x}, \qquad \hat{\mu}_2 = \bar{y}, \qquad \hat{\sigma}^2 = \frac{1}{n}\left[\sum(x_i - \bar{x})^2 + \sum(y_i - \bar{y})^2\right].$$

As a result,

$$L(x, y \mid \hat{\theta}) = \frac{e^{-n/2}}{(2\pi)^{n/2}}\left[\frac{\sum(x_i - \bar{x})^2 + \sum(y_i - \bar{y})^2}{n}\right]^{-n/2}.$$

Similarly, the estimates for (34) turn out to be

$$\hat{\mu} = \frac{1}{n}\left[\sum x_i + \sum y_i\right], \qquad \hat{\sigma}^2 = \frac{1}{n}\left[\sum(x_i - \hat{\mu})^2 + \sum(y_i - \hat{\mu})^2\right].$$

Hence,

$$L(x, y \mid \hat{\theta}_0) = \frac{e^{-n/2}}{(2\pi)^{n/2}}\left[\frac{\sum(x_i - \hat{\mu})^2 + \sum(y_i - \hat{\mu})^2}{n}\right]^{-n/2}.$$

The same techniques that were used in the preceding illustration will reduce the critical region $\lambda \leq \lambda_0$ for this likelihood ratio test to an equivalent critical region which can be expressed as

$$(35) \quad \frac{\sum(x_i - \hat{\mu})^2 + \sum(y_i - \hat{\mu})^2}{\sum(x_i - \bar{x})^2 + \sum(y_i - \bar{y})^2} \geq k.$$

As before, algebraic techniques will reduce this critical region to one that can be based on a Student t variable. Thus,

$$\sum (x_i - \hat{\mu})^2 + \sum (y_i - \hat{\mu})^2 = \sum (x_i - \bar{x})^2 + \sum (y_i - \bar{y})^2$$
$$+ n_1(\bar{x} - \hat{\mu})^2 + n_2(\bar{y} - \hat{\mu})^2.$$

But $\hat{\mu} = \dfrac{n_1\bar{x} + n_2\bar{y}}{n}$; hence $\bar{x} - \hat{\mu} = \dfrac{n_2(\bar{x} - \bar{y})}{n}$ and $\bar{y} - \hat{\mu} = \dfrac{n_1(\bar{y} - \bar{x})}{n}$.

When these results are applied to (35), it will reduce to

$$\frac{n_1 n_2 (\bar{x} - \bar{y})^2}{n[\sum (x_i - \bar{x})^2 + \sum (y_i - \bar{y})^2]} \geq k - 1.$$

This is equivalent to the inequality

(36) $$\frac{n_1 n_2 (\bar{x} - \bar{y})^2 (n_1 + n_2 - 2)}{n[\sum (x_i - \bar{x})^2 + \sum (y_i - \bar{y})^2]} \geq (k - 1)(n_1 + n_2 - 2).$$

For the purpose of showing that the left side is the square of a Student t variable, let

$$u = \frac{\overline{X} - \overline{Y}}{\sigma\sqrt{\dfrac{1}{n_1} + \dfrac{1}{n_2}}}.$$

Since $\overline{X} - \overline{Y}$ is a normal variable with $E(\overline{X} - \overline{Y}) = \mu_1 - \mu_2 = 0$ under H_0 and $V(\overline{X} - \overline{Y}) = V(\overline{X}) + V(\overline{Y}) = \dfrac{\sigma^2}{n_1} + \dfrac{\sigma^2}{n_2} = \sigma^2\left(\dfrac{1}{n_1} + \dfrac{1}{n_2}\right)$, it follows that u qualifies as a standard normal variable. Furthermore, since $\dfrac{\sum (X_i - \overline{X})^2}{\sigma^2}$ and $\dfrac{\sum (Y_i - \overline{Y})^2}{\sigma^2}$ are independent chi-square variables with $n_1 - 1$ and $n_2 - 1$ degrees of freedom, respectively, it follows that their sum, namely $\dfrac{[\sum (X_i - \overline{X})^2 + \sum (Y_i - \overline{Y})^2]}{\sigma^2}$, is a chi-square variable with $n_1 + n_2 - 2$ degrees of freedom, because we know from Volume I that the sum of two independent chi-square variables is a chi-square variable with degrees of freedom that add. Finally, since it was assumed earlier that $\overline{X}$ and $\sum (X_i - \overline{X})^2$ are independent random variables, it follows from the independence of X and Y that the preceding chi-square variable is independent of the variable u, and therefore from definition (31) that

(37) $$\frac{(\overline{X} - \overline{Y})\sqrt{n_1 + n_2 - 2}}{\sqrt{\dfrac{1}{n_1} + \dfrac{1}{n_2}}\sqrt{\sum (X_i - \overline{X})^2 + \sum (Y_i - \overline{Y})^2}} = t$$

possesses a Student t distribution with $n_1 + n_2 - 2$ degrees of freedom. As in the case of (32), the unknown σ that occurs in our ratio cancels out.

A comparison of (36) and (37) will show that our likelihood ratio test is equivalent to a test using as critical region equal tails of the Student t distribution with $n_1 + n_2 - 2$ degrees of freedom, where t is given by (37).

As a numerical illustration, consider the following data on the gain in weight of twenty rats, half of which received their protein from raw peanuts and half from roasted peanuts. The problem is to test whether the roasting had any effect as far as gain in weight is concerned.

Raw	62	56	61	58	60	44	56	60	56	63
Roasted	53	51	62	55	59	56	61	54	47	57

If the first and second rows of values are denoted by x and y, respectively, calculations will give

$$\bar{x} = 57.6, \qquad \bar{y} = 55.5, \qquad \sum (x_i - \bar{x})^2 = 264, \qquad \sum (y_i - \bar{y})^2 = 189.$$

Hence, applying (37),

$$t = \frac{2.1\sqrt{18}}{\sqrt{\dfrac{1}{10} + \dfrac{1}{10}}\sqrt{453}} = .94.$$

From Table VI in the appendix, the .05 critical point for the t distribution based on 18 degrees of freedom and assuming a two-sided critical region is given by $t_0 = 2.101$. Thus, the hypothesis $H_0 : \mu_1 = \mu_2$ is accepted, and on the basis of these experimental data there is no reason for believing that the roasting affected the gain producing properties of the peanuts.

3.5. Goodness of fit tests

The likelihood ratio test can be applied to a problem that is of much practical interest and which occurs in many different forms. In its simplest abstract form, we are given k cells into one of which the outcome of an experiment must fall. We shall let p_i, $i = 1, \ldots, k$, denote the probability that the experiment will produce an outcome falling in the ith cell, and let n_i, $i = 1, \ldots, k$, denote the number of outcomes that fell in the ith cell in a total of $n = \sum_{i=1}^{k} n_i$ experiments. A physical model to represent this mathematical formulation is the carnival game of rolling a ball up an inclined plane into compartments that are assigned different point values, with the contestant having a fixed number of balls to roll. The point values can be thought of as weights that correspond inversely to the probabilities of landing in the various compartments.

Since any given cell probability p_i can be thought of as the binomial probability of success in a single trial of an experiment and since n trials are to occur, the expected number of successes for the ith cell will be the mean of the corresponding binomial variable, namely np_i. It is convenient to represent the preceding notation and problem in the following form

o_i	n_1	n_2	$\cdots$	n_k
e_i	np_1	np_2	$\cdots$	np_k

Here the row labeled o_i represents the observed frequencies, and the row labeled e_i represents the expected frequencies in a total of n experiments.

The basic problem is the following one. How can we test to see whether a set of experimental outcomes is compatible with the outcomes expected on the basis of the probabilities that were postulated for the cells? Formally, we wish to test the hypothesis

$$H_0: p_i = \pi_i, \qquad i = 1, \ldots, k,$$

where the π's are the postulated values of the cell probabilities. This is actually a simple hypothesis; however, since we do not ordinarily have a specific alternative hypothesis in mind, we shall employ a likelihood ratio test here. Now the likelihood function for a discrete variable is the probability of obtaining the observed sample values in the order in which they were obtained; therefore

$$L(x \mid \Theta) = p_1^{n_1} \cdots p_k^{n_k}.$$

Here x represents the n sample values that gave rise to the cell frequencies $n_1, \ldots, n_k$. Since $\sum_1^k p_i = 1$, only $k - 1$ of the p's are independent parameters; therefore, in finding the maximum likelihood estimates of the p's, it is advisable to replace p_k by $1 - \sum_1^{k-1} p_i$. Using calculus methods, we first calculate partial derivatives and set them equal to zero to obtain critical points. Taking logarithms and differentiating with respect to p_r, where r is any index, we obtain the equations

$$\log L = \sum_{i=1}^{k-1} n_i \log p_i + n_k \log \left(1 - \sum_{i=1}^{k-1} p_i \right),$$

and

$$\frac{\partial \log L}{\partial p_r} = \frac{n_r}{p_r} - \frac{n_k}{1 - \sum\limits_{i=1}^{k-1} p_i} = \frac{n_r}{p_r} - \frac{n_k}{p_k}, \qquad r = 1, \ldots, k - 1.$$

Setting these derivatives equal to 0 and adding a kth equation will give

$$p_r = \frac{p_k}{n_k} n_r, \qquad r = 1, \ldots, k.$$

Summing both sides with respect to r will produce 1 on the left side and $p_k n/n_k$ on the right side; hence $p_k = n_k/n$. As a result we obtain the solution

$$\hat{p}_i = \frac{n_i}{n}, \qquad i = 1, \ldots, k.$$

It can be shown that these calculus estimates, which are the natural estimates here, do maximize the likelihood function.

Since

$$L(x \mid \Theta_0) = \pi_1^{n_1} \cdots \pi_k^{n_k},$$

there are no unspecified parameters here, and therefore no estimates are needed. The ratio (25) then becomes

$$\lambda = \frac{\pi_1^{N_1} \cdots \pi_k^{N_k}}{\left(\dfrac{N_1}{n}\right)^{N_1} \cdots \left(\dfrac{N_k}{n}\right)^{N_k}} = \left(\frac{n\pi_1}{N_1}\right)^{N_1} \cdots \left(\frac{n\pi_k}{N_k}\right)^{N_k},$$

where N_i denotes the random variable of which n_i is the observed value. Unfortunately, finding the distribution of λ is a hopeless task here; therefore we shall rely upon using its asymptotic distribution. Since $-2 \log \lambda$ possesses an asymptotic chi-square distribution with degrees of freedom equal to the difference in the number of independent parameters unspecified under Θ and Θ_0, it follows that the proper number of degrees of freedom here is $k - 1$. Thus, our test reduces to treating

$$(38) \qquad -2 \sum_{i=1}^{k} N_i \log \left(\frac{n\pi_i}{N_i}\right)$$

as a chi-square variable with $k - 1$ degrees of freedom and choosing as critical region the region $\chi^2 \geq \chi_0^2$ where $P(\chi^2 \geq \chi_0^2) = \alpha$. The right tail is chosen because, as before, $\lambda \leq \lambda_0$ is equivalent to $\log \lambda \leq \log \lambda_0$, which in turn is equivalent to $-2 \log \lambda \geq -2 \log \lambda_0$ and hence to $\chi^2 \geq -2 \log \lambda_0$.

Before likelihood ratio tests evolved, a somewhat different approach to testing whether a set of observed frequencies is compatible with a set of expected frequencies was introduced. It consisted in treating the random variable

$$(39) \qquad \sum_{i=1}^{k} \frac{(N_i - e_i)^2}{e_i}$$

as a χ^2 variable with $k - 1$ degrees of freedom and using the right tail of the χ^2 distribution as critical region. A mathematical demonstration can be given to show that the random variable defined by (39) possesses an asymptotic χ^2 distribution with $k - 1$ degrees of freedom, but it is quite lengthy and will not be treated here. If we accept the fact that the likelihood ratio statistic (38) possesses an asymptotic χ^2 distribution, then it is not too difficult to show that (39) will possess this same asymptotic distribution. The demonstration consists in expanding the logarithms in (38) in an appropriate manner, showing that only the leading terms are important, and finally showing that when this is done we will obtain

$$-2 \log \lambda = \sum_{i=1}^{k} \frac{(N_i - n\pi_i)^2}{n\pi_i} + R_n,$$

where $R_n \to 0$ in probability as $n \to \infty$. Thus, asymptotically there is no difference between these tests. Because of long tradition and usage, the statistic (39) is customarily used in goodness of fit problems. It is simpler to apply than (38), because (38) requires calculations involving natural logarithms.

As an illustration, consider the following data shown in the first row of Table 2 on the number of individuals in a sample of 770 who possessed various blood types. The sample was taken from a race of people for which it is postulated that the proportions of the various types are .16, .48, .20, and .16, respectively. We wish to test the hypothesis that these are correct values for the cell probabilities.

Table 2

o_i	180	360	132	98
e_i	123	370	154	123

The expected values shown in Table 2 were obtained by multiplying 770 by the postulated probabilities, but rounding off to the nearest integer. First, let us apply (39). Here

$$\chi^2 = \frac{(180 - 123)^2}{123} + \frac{(360 - 370)^2}{370} + \frac{(132 - 154)^2}{154}$$
$$+ \frac{(98 - 123)^2}{123} \doteq 35.$$

Since from Table V in the appendix $P(\chi^2 > 7.8) = .05$ for three degrees of freedom, the value of $\chi^2 = 35$ is far out in the critical region, and therefore the hypothesis $H_0: p_1 = .16, p_2 = .48, p_3 = .20, p_4 = .16$ is easily rejected. Next, let us work the problem by using (38). Here

$$\chi^2 = -2 \left[180 \log \frac{123}{180} + 360 \log \frac{370}{360} + 132 \log \frac{154}{132} + 98 \log \frac{123}{98} \right]$$
$$= -2[-70.0 + 9.7 + 20.3 + 22.2] \doteq 36.$$

Thus, there is no essential difference in the χ^2 values of these two asymptotically equivalent tests. Since n is large here, this should come as no surprise.

Because the classical χ^2 test is easier to apply than our unmodified likelihood ratio test and since both of them are only large sample approximations to an exact test, we shall use the classical version hereafter.

As a second illustration, consider the data in Table 3 on the number of α-particles radiated from a plate during a five-second interval for a total of 500 such intervals.

Table 3

i	0	1	2	3	4	5	6	7	8	9	10	11
f_i	11	40	74	100	103	79	50	26	9	5	2	1

Here f_i denotes the number of five-second time intervals in which the number of α-particles radiated was equal to i. The problem is to test whether these α-particle counts behave like a random sample of 500 from a Poisson distribution.

This problem differs from the preceding one in that we are not given a set of postulated cell probabilities to test. If we knew the mean μ of the Poisson distribution that is being postulated here, the problem could be formulated as one of testing the hypothesis

$$H_0: p_i = \frac{e^{-\mu}\mu^i}{i!}, \qquad i = 0, 1, \ldots.$$

A difficulty that arises with this formulation, however, even if μ is known, is that theoretically there is an infinite number of cells. This can be overcome by combining all cells beyond a certain stage into one cell. Now it is known that the χ^2 test may not be reliable if any of the cell frequencies is smaller than 4 or 5. Thus, we can circumvent the difficulty of too many cells by combining the cells for $i \geq 9$ into one cell, which means that we will have $k = 10$ cells.

The first difficulty, of not knowing the value of μ, can also be circumvented by means of a theoretical result which will permit us to replace μ by its sample estimate. This theoretical result, which is quite general, may be stated as a property of the χ^2 test as follows.

Property 2   *The χ^2 test is applicable when the cell probabilities depend upon unknown parameters, provided the unknown parameters are replaced by their maximum likelihood estimates and provided one degree of freedom is deducted for each such parameter being estimated.*

This property follows from the asymptotic property of the likelihood ratio test as it applies to testing a set of cell probabilities corresponding to the likelihood function $L = p_1^{n_1} \cdots p_k^{n_k}$ when unknown parameters in L are replaced by their maximum likelihood estimates.

Theoretically, this property requires that the maximum likelihood estimates of the unknown parameters be those that maximize the likelihood function $p_1^{n_1} \cdots p_k^{n_k}$ after any contemplated amalgamations have taken place. These estimates may differ somewhat from the ordinary maximum likelihood estimates that we have calculated previously, if k is small and considerable amalgamation takes place. However, for large sample problems such as the one we are considering, the difference in the estimates is so slight that it can be ignored.

The use of Property 2 will permit us to solve the preceding problem. Calculations based on Table 3 will show that $\bar{x} = 3.85$; therefore we substitute that value for μ in the formula for p_i. The correct estimate based on maximizing $p_1^{n_1} \cdots p_k^{n_k}$ for Table 4 with $k = 10$ also turns out to be 3.85, correct to two decimals; therefore the simpler traditional estimate is certainly appropriate here. The expected frequencies e_i for the various cells are then obtained from $500\, p_i$. Thus,

$$(40) \qquad e_i = 500\, \frac{e^{-3.85}(3.85)^i}{i!}, \qquad i = 0, 1, \ldots.$$

Calculations based on this formula yielded the values shown in Table 4.

Table 4

i	0	1	2	3	4	5	6	7	8	≥ 9
o_i	11	40	74	100	103	79	50	26	9	8
e_i	10.6	41.0	78.9	101.2	97.4	75.0	48.1	26.5	12.8	8.5

The computation of χ^2 based on (39) then yielded the value $\chi^2 = 2.8$.

Since the amalgamation of cells made $k = 10$ and since μ is the only parameter that was estimated from the data, our χ^2 variable will have 8 degrees of freedom. From Table V in the appendix the 5 percent critical point on the χ^2 distribution corresponding to 8 degrees of freedom is $\chi_0^2 = 15.5$; hence H_0 is accepted here. This implies that as far as the distribution of counts is concerned, they do behave like random samples of a Poisson variable

As a third application, consider the problem of testing whether two variables that have been classified by means of a two-way table are independent random variables. Table 5 is an illustration of this type of problem. The data for the table were obtained from a random sample of 300 college students. The students were classified with respect to the size of the high school from which they graduated and with respect to their freshman year grade point average. The problem is to test whether there is any relationship between these two classified variables.

Table 5

High School

		Small	Medium	Large
Record	Above 2.5 average	18	51	46
	Below 2.5 average	42	79	64

For a more general version of such a two-way table, which is customarily called a *contingency table*, assume that the classification uses r rows and c columns, and let p_{ij} denote the probability that an individual selected at random from the population under consideration will fall in the cell corresponding to the ith row and jth column. Further, let $p_{i.} = \sum_{j=1}^{c} p_{ij}$ denote the probability of falling in the ith row. Similarly, let $p_{.j} = \sum_{i=1}^{r} p_{ij}$ be the corresponding probability for the jth column.

The problem of testing whether the two classified variables are independent may now be formalized as the problem of testing the hypothesis

$$H_0 : p_{ij} = p_{i.} p_{.j}, \qquad \begin{cases} i = 1, \ldots, r \\ j = 1, \ldots, c \,. \end{cases}$$

The likelihood function is the same as before, except that a more complicated notation is needed to express it. Thus,

$$L(x \mid \Theta) = \prod_{i=1}^{r} \prod_{j=1}^{c} p_{ij}^{n_{ij}},$$

where n_{ij} is the observed frequency in the cell for which p_{ij} is the probability. When H_0 is true, L becomes

$$L(x \mid \Theta_0) = \prod_{i=1}^{r} \prod_{j=1}^{c} (p_{i.} p_{.j})^{n_{ij}}.$$

Since the $p_{i.}$ and $p_{.j}$ are unknown parameters, the expected values $e_{ij} = np_{ij} = np_{i.} p_{.j}$ needed for the χ^2 test are unknown. It therefore follows from Property 2 that they must be replaced by their maximum likelihood estimates before the test can be performed. Standard calculus methods based on calculating the proper partial derivatives will show that these estimates are given by the formulas

$$\hat{p}_{i.} = \frac{\sum_{j=1}^{c} n_{ij}}{n} = \frac{n_{i.}}{n} \quad \text{and} \quad \hat{p}_{.j} = \frac{\sum_{i=1}^{r} n_{ij}}{n} = \frac{n_{.j}}{n}.$$

In calculating the partial derivatives it is necessary to realize that $\sum_{i=1}^{r} p_{i.} = 1$ and $\sum_{j=1}^{c} p_{.j} = 1$, and therefore that only $r + c - 2$ of these parameters may be treated as independent calculus variables.

Application of the preceding results to (39) yields the general formula

$$\chi^2 = \sum_{i=1}^{r} \sum_{j=1}^{c} \frac{\left(n_{ij} - \dfrac{n_{i.}n_{.j}}{n}\right)^2}{\dfrac{n_{i.}n_{.j}}{n}}.$$

Since $k - 1 = rc - 1$ and since $r + c - 2$ parameters are being replaced by their maximum likelihood estimates, it follows from Property 2 that the number of degrees of freedom for the test is given by

$$v = rc - 1 - (r + c - 2) = (r - 1)(c - 1).$$

This general formula will now be applied to the data of Table 5. Calculations based on those data give the expected frequencies, namely $n_{i.}n_{.j}/n$, shown in parentheses in Table 6. Application of the preceding formula for χ^2 then gives

$$\chi^2 = \frac{(18 - 23)^2}{23} + \frac{(51 - 50)^2}{50} + \frac{(46 - 42)^2}{42} + \frac{(42 - 37)^2}{37}$$

$$+ \frac{(79 - 80)^2}{80} + \frac{(64 - 68)^2}{68} = 2.4.$$

For $\alpha = .05$ and $v = 2$, Table V shows that $\chi_0^2 = 6.0$; therefore the hypothesis that these two variables are independent random variables is accepted. There is no evidence here that the size of high school is related to college grades.

Table 6

High School

		Small	Medium	Large		
Record	Above 2.5 average	18 (23)	51 (50)	46 (42)	115	$(n_1.)$
	Below 2.5 average	42 (37)	79 (80)	64 (68)	185	$(n_2.)$
		60 $(n_{.1})$	130 $(n_{.2})$	110 $(n_{.3})$	300	(n)

The two preceding devices of combining cells and estimating unknown parameters permit the χ^2 test to be applied to a large variety of practical problems and thereby make it one of the most popular and useful of all statistical tools.

BAYESIAN METHODS

3.6. Testing simple hypotheses

We shall now assume that in addition to being given a sample of size n of the variable X whose probability density is $f(x \mid \theta)$, we are also given prior information concerning θ. Since we shall be testing $H_0: \theta = \theta_0$ against $H_1: \theta = \theta_1$, the only information needed is contained in the prior probabilities $\pi(\theta_0)$ and $\pi(\theta_1)$ that θ will have the values θ_0 and θ_1, respectively. In the present problem it is assumed that θ_0 and θ_1 are the only conceivable values of θ; therefore it must be true that $\pi(\theta_1) = 1 - \pi(\theta_0)$. We shall, however, use the notation $\pi_0 = \pi(\theta_0)$ and $\pi_1 = \pi(\theta_1)$ because it gives neater formulas.

Since the basis for making decisions is now the mean risk, it is necessary to calculate the expected value of the risk function given by (1), and then determine the decision function d that minimizes this expected value. From (1) we obtain

$$(41) \qquad r(\pi, d) = E\mathscr{R}(\theta, d) = \pi_0 \mathscr{R}(\theta_0, d) + \pi_1 \mathscr{R}(\theta_1, d)$$

$$= \pi_0 P(X \in A_1 \mid \theta_0) + \pi_1 P(X \in A_0 \mid \theta_1).$$

As you may recall, the Neyman-Pearson Lemma enabled us to find a best critical region by considering critical regions for which $P(X \in A_1 \mid \theta_0)$ had a fixed value and then choosing one for which $P(X \in A_0 \mid \theta_1)$ was a minimum. Here also it would be desirable to choose a critical region that minimizes $P(X \in A_0 \mid \theta_1)$ among those for which $P(X \in A_1 \mid \theta_0)$ has a fixed value. The difference is that we no longer have a fixed value of α with which to specify the value of $P(X \in A_1 \mid \theta_0)$. Instead it is necessary to allow this probability to assume any value which together with the corresponding value of $P(X \in A_0 \mid \theta_1)$ will minimize $r(\pi, d)$. Except for this lack of flexibility in assigning the value of $P(X \in A_1 \mid \theta_0)$ in advance, the two problems are so much alike that it would be surprising if there were not a theorem similar to the Neyman-Pearson Lemma for finding a best critical region in the Bayesian sense. Such a theorem does exist in the following form.

Theorem 1 *A Bayes solution to testing $H_0: \theta = \theta_0$ against $H_1: \theta = \theta_1$ for the probability density $f(x \mid \theta)$ on the basis of a random* *sample of size n of the variable X and the prior probabilities π_0 and π_1 for the values θ_0 and θ_1, is given by choosing as critical region those sample points where*

$$(42) \qquad \frac{\prod_{i=1}^{n} f(x_i \mid \theta_1)}{\prod_{i=1}^{n} f(x_i \mid \theta_0)} \geq \frac{\pi_0}{\pi_1}.$$

This theorem is a special case of a more general theorem that is proved in the next section; therefore a proof will not be given here.

As an illustration, consider the first example following the proof of the Neyman-Pearson Lemma in which the problem was to test $H_0: \mu = \mu_0$ against $H_1: \mu = \mu_1 > \mu_0$ for a normal density with known variance σ^2. We shall assume that our prior probabilities are π_0 and π_1, respectively. Since $\dfrac{L_1}{L_0}$ was calculated previously in (9), a direct application of (42) will yield the inequality

$$\exp \left[\frac{\mu_1 - \mu_0}{\sigma^2} \sum x_i + \frac{n(\mu_0^2 - \mu_1^2)}{2\sigma^2} \right] \geq \frac{\pi_0}{\pi_1}.$$

Taking logarithms will yield an equivalent inequality which reduces to

$$(43) \qquad \bar{x} \geq \frac{\mu_0 + \mu_1}{2} + \frac{\sigma^2}{n} \frac{\log (\pi_0/\pi_1)}{\mu_1 - \mu_0}.$$

From this result it will be observed that when $\pi_0 = \pi_1$, the borderline for deciding in favor of $\mu = \mu_0$ against $\mu = \mu_1$ is a value of $\bar{x}$ half way between μ_0 and μ_1. However, if $\pi_0 > \pi_1$ the borderline will be to the right of this half-way point, with the extent of this shift depending upon the magnitudes of the constants σ^2, $\mu_1 - \mu_0$, and $\log (\pi_0/\pi_1)$, and upon the size of the sample. Shifting the borderline to the right will decrease the size of the critical region, and therefore decisions in favor of $\mu = \mu_0$ will become more numerous. This is as it should be, because $\pi_0 > \pi_1$ implies that H_0 is more likely to be true than H_1. As n becomes large, however, the second term of (43) approaches zero regardless of whether $\pi_0 > \pi_1$ or $\pi_0 < \pi_1$, which means that the information supplied by the knowledge of π_0 and π_1 decreases in importance as the sample size increases.

It is also interesting to observe that our choice of loss function results in the mean risk becoming the probability of making an incorrect decision. This fact follows directly from Formula (41). The Bayes solution given by (42) is therefore a solution that minimizes the probability of making an incorrect decision.

3.7. Multiple decision problems

Of the three classes of statistical problems enumerated in Chapter 1, only the multiple decision problem has not been studied thus far. We shall discuss it here from a Bayesian point of view because it occurs frequently in that type of setting and because it serves as a generalization of the problem discussed in the preceding section.

Suppose a sample of size n is taken of a random variable X whose probability density $f(x \mid \theta)$ depends upon a parameter θ which can have any one of k values $\theta_1, \ldots, \theta_k$, and that the particular value of θ is unknown. The problem then is to devise a test for deciding which is the correct value of θ. For $k = 2$, this is merely the problem discussed in the preceding section. It is assumed here that the prior probabilities $\pi(\theta_i) = \pi_i$, $i = 1, \ldots, k$, are given, that we are using the loss function

$$\mathcal{L}(\theta_i, a) = \begin{cases} 0, & \text{if the correct decision is made} \\ 1, & \text{if an incorrect decision is made} \end{cases},$$

and that we wish to minimize the mean risk.

By analogy with the $k = 2$ problem, we should expect the likelihood ratio to play a dominant role here just as it has in all the preceding tests. Furthermore, in view of (42) we should expect to place a sample point in the acceptance region for $\theta = \theta_i$ if $\pi_i L_i \geq \pi_j L_j$ for all $j \neq i$. These beliefs are justified in the following theorem, which gives a method for constructing a Bayes solution for the problem.

Theorem 2   *A Bayes solution to the multiple decision problem is given by choosing as acceptance region for the hypothesis H_i: $\theta = \theta_i$, $i = 1, \ldots, k$, that part of sample space where*

$$(44) \quad \pi_i \prod_{s=1}^{n} f(x_s \mid \theta_i) \geq \pi_j \prod_{s=1}^{n} f(x_s \mid \theta_j) \quad \text{for all} \quad j \neq i,$$

and where any point lying in more than one such region is assigned to one of them.

Proof.   Let A_i denote the acceptance region for $\theta = \theta_i$ determined by the theorem and any convention concerning non-unique points. A unique set of acceptance regions could be obtained, for example, by the simple device of assigning any non-uniquely determined point to the acceptance region with the smallest index. Let B_i denote the corresponding acceptance region given by any other partitioning of sample space. We shall introduce the two indicator functions

$$\varphi_i(x) = \begin{cases} 1 & \text{if } x \in A_i \\ 0 & \text{otherwise} \end{cases} \quad \text{and} \quad \psi_i(x) = \begin{cases} 1 & \text{if } x \in B_i \\ 0 & \text{otherwise} \end{cases}.$$

Let d and δ denote the corresponding decision functions. If for brevity of notation we write $f_i(x)$ for $\prod_{s=1}^{n} f(x_s \mid \theta_i)$ and use condensed integration notation, then for X a continuous variable

$$r(\pi, \delta) - r(\pi, d) = \sum_{j=1}^{k} \pi_j \mathscr{R}(\theta_j, \delta) - \sum_{i=1}^{k} \pi_i \mathscr{R}(\theta_i, d)$$

$$= \sum_{j=1}^{k} \pi_j \int \mathscr{L}(\theta_j, \delta) f_j(x) \, dx$$

$$- \sum_{i=1}^{k} \pi_i \int \mathscr{L}(\theta_i, d) f_i(x) \, dx.$$

We are assuming that our loss function will have the value 0 for any point in the acceptance region and have the value 1 for all other points; hence by letting S denote the entire sample space, we can write

$$r(\pi, \delta) - r(\pi, d) = \sum_{j=1}^{k} \pi_j \int_{S-B_j} f_j(x) \, dx - \sum_{i=1}^{k} \pi_i \int_{S-A_i} f_i(x) \, dx.$$

Now the sum of these integrals over S, weighted with the π's, will cancel from the two expressions; therefore after eliminating them we will obtain

$$r(\pi, \delta) - r(\pi, d) = \sum_{i=1}^{k} \pi_i \int_{A_i} f_i(x) \, dx - \sum_{j=1}^{k} \pi_j \int_{B_j} f_j(x) \, dx.$$

The introduction of our indicator functions will enable us to integrate over the entire sample space. Thus, we can write

$$r(\pi, \delta) - r(\pi, d) = \sum_{i=1}^{k} \pi_i \int_{S} \varphi_i(x) f_i(x) \, dx - \sum_{j=1}^{k} \pi_j \int_{S} \psi_j(x) f_j(x) \, dx.$$

Since $\sum_{i=1}^{k} \varphi_i(x) = 1$ and $\sum_{j=1}^{k} \psi_j(x) = 1$ hold for all x, we may insert such sums as factors in the integrands of the preceding integrals to obtain

$$r(\pi, \delta) - r(\pi, d) = \sum_{i=1}^{k} \pi_i \int_{S} \varphi_i(x) f_i(x) \sum_{j=1}^{k} \psi_j(x) \, dx$$

$$- \sum_{j=1}^{k} \pi_j \int_{S} \psi_j(x) f_j(x) \sum_{i=1}^{k} \varphi_i(x) \, dx$$

$$= \sum_{i=1}^{k} \sum_{j=1}^{k} \int_{S} \varphi_i(x) \psi_j(x) [\pi_i f_i(x) - \pi_j f_j(x)] \, dx.$$

But from the definition of A_i as given by (44), it follows that $\varphi_i(x) = 0$ for any point x where the quantity in brackets is negative; therefore the integrand is nonnegative. This holds for every i and j; consequently it follows that

$$r(\pi, \delta) \geq r(\pi, d),$$

and hence that d is a Bayes solution to the multiple decision problem. ∎

It should be noted that Theorem 1 is a special case of this theorem. As in the case of that theorem, the mean risk gives the probability of making an incorrect decision, so that the Bayes solution given by (44) minimizes the probability of making an incorrect decision in the multiple decision problem.

As an illustration, let $f(x \mid \theta) = \dfrac{e^{-x/\theta}}{\theta}$, $x > 0$; let $\theta_1 = 1$, $\theta_2 = 2$, $\theta_3 = 4$; and let the prior probabilities be given by $\pi_1 = 1/4$, $\pi_2 = 1/2$, $\pi_3 = 1/4$. Suppose a single observation is to be used to make a decision. The acceptance regions will then be given by those values of x that satisfy the proper inequalities. Thus, from (44), A_1 is determined by the inequalities

$$\frac{e^{-x}}{4} \geq \frac{e^{-x/2}}{4} \quad \text{and} \quad \frac{e^{-x}}{4} \geq \frac{e^{-x/4}}{16}.$$

The first inequality is equivalent to $e^{-x/2} \geq 1$ which is possible only for $x = 0$. Since the second inequality is satisfied for $x = 0$, A_1 can consist of at most the point $x = 0$. The region A_2 is determined by the inequalities

$$\frac{e^{-x/2}}{4} \geq \frac{e^{-x}}{4} \quad \text{and} \quad \frac{e^{-x/2}}{4} \geq \frac{e^{-x/4}}{16}.$$

It is clear that the first inequality is true for all $x \geq 0$. By the use of logarithms, the second inequality will be seen to be equivalent to $x \leq 4 \log 4$. Thus, if we assign the boundary point $x = 0$ to A_1 rather than A_2, the region A_2 will be the interval $0 < x \leq 4 \log 4$. The region A_3 will then be what is left of the positive x axis, provided the boundary point $x = 4 \log 4$ is assigned to A_2 rather than A_3. Thus, a Bayes solution to this problem is given by choosing as acceptance regions for the respective hypotheses $H_1: \theta = 1$, $H_2: \theta = 2$, and $H_3: \theta = 4$, the point $x = 0$, the interval $0 < x \leq 4 \log 4$, and the interval $x > 4 \log 4$. If we had assigned the boundary point $x = 0$ to A_2 rather than to A_1, our Bayes solution would tell us to never decide in favor of H_1 when only a single sample is taken. Since X is a continuous random variable, the probability of obtaining $X = 0$ is zero; consequently the earlier choice of A_1 as the point $x = 0$ still says that we will never decide in favor of H_1. For larger samples this seeming paradox will not occur.

Exercises

1 Let H_0 be the hypothesis that $p = 1/2$ and H_1 the hypothesis that $p = 2/3$, where p is the probability of success in an experiment. The experiment is to be performed twice. If it is agreed to accept H_0 if, and only if, two successes are obtained, what are the values of α and β?

2 Given $f(x \mid \theta) = 1/\theta, 0 \le x \le \theta$, and given the hypotheses $H_0: \theta = 1$ and $H_1: \theta = 2$, suppose a single observed value of X is to be taken.
(a) If the critical region is chosen to be the interval $X > .5$, what are the values of α and β?
(b) What would those values become if $X > 1.5$ were chosen as the critical region?

3 A box is known to contain either 3 red and 5 black balls or 5 red and 3 black balls. Three balls are to be drawn from the box; and, if less than three red balls are obtained, the decision will be made that the box contains 3 red and 5 black balls. Calculate the values of α and β.

4 Let X be a discrete variable whose density values under H_0 and H_1 are given by the following table.
(a) List all the critical regions of size $\alpha = .10$.
(b) Among the critical regions listed in (a), find the one with the smallest value of β.

x	1	2	3	4	5	6	7
$f(x \mid H_0)$	.01	.02	.03	.05	.05	.07	.77
$f(x \mid H_1)$	.03	.09	.10	.10	.20	.18	.30

5 Suppose you are testing $H_0: p = 1/2$ against $H_1: p = 2/3$ for a binomial variable X with $n = 3$. What values of X would you assign to the critical region if you wish to have $\alpha \le 1/8$ and you wish to minimize β corresponding to the value of α selected?

6 A box is known to contain 8 worthless tickets and either 1 or 2 prize winning tickets. Let H_0 and H_1 denote these two possibilities. To test H_0, tickets will be drawn one at a time from the box until a prize winning ticket is obtained. Let X denote the number of drawings required, and calculate $f(x \mid H_0)$ and $f(x \mid H_1)$. Choose as critical region for the test those points for which $X \le 5$. Calculate α and β.

7 Consider the hypotheses $H_0: p = 1/2$ and $H_1: p = 1$ for binomial X for which $n = 2$. List all possible critical regions for which $\alpha \le 1/2$. Which of these critical regions minimizes $\alpha + \beta$?

8 Determine the nature of the best critical region, based on a sample of size n, for testing $H_0: \theta = \theta_0$ against $H_1: \theta = \theta_1 < \theta_0$ if $f(x \mid \theta) = (1 + \theta)x^\theta, 0 \le x \le 1, \theta > 0$.

9 Determine the nature of the best critical region, based on a sample of size n, for testing $H_0: \theta = \theta_0$ against $H_1: \theta = \theta_1 < \theta_0$ if $f(x \mid \theta) = \theta e^{-\theta x}, x > 0, \theta > 0$.

10 Determine the nature of the best critical region, based on a sample of size n, for testing $H_0: \theta = \theta_0$ against $H_1: \theta = \theta_1$ if $f(x \mid \theta) = c(\theta)h(x)e^{a(\theta)b(x)}$, where $c, h, a,$ and b are functions of their arguments.

11 Given $f(x \mid q) = (1 - q)q^{x-1}$, $x = 1, 2, \ldots$, find a best test for testing $H_0: q = 3/4$ against $H_1: q = 2/3$ based on a single observed value of X. Choose a convenient value of α that satisfies $\alpha \leq 1/2$.

12 Given $\bar{x} = 28$ for a sample of 50 of a normal variable for which $\sigma = 5$, test $H_0: \mu = 30$ against $H_1: \mu = 29$ with $\alpha = .05$.

13 Given $\sum x_i^2 = 1600$ for a sample of 20 of a normal variable for which $\mu = 0$, test $H_0: \sigma = 6$ against $H_1: \sigma = 9$ with $\alpha = .05$.

14 Given $\sum (x_i - 20)^2 = 1000$ for a sample of 30 of a normal variable for which $\mu = 20$, test $H_0: \sigma = 8$ against $H_1: \sigma = 6$ with $\alpha = .025$.

15 Take a sample of 25 random normal numbers from Table III in the appendix, and use them to test the hypothesis $H_0: \mu = 0$ against $H_1: \mu = .2$. Use the fact that $\sigma = 1$ and choose $\alpha = .10$.

16 Test $H_0: p = .5$ against $H_1: p = .6$ if 80 trials of an experiment produced 50 successes and if $\alpha = .10$. Use the normal approximation for binomial X.

17 Test $H_0: \mu = 10$ against $H_1: \mu = 12$ if 10 experimental values of a Poisson variable X yielded $\sum x_i = 116$ and if $\alpha = .10$. Use the normal approximation for Poisson X.

18 Test $H_0: p_1 = p_2$ against $H_1: p_1 \neq p_2$ if 120 trials produced 60 successes for the first experiment and 160 trials produced 96 successes for the second experiment, where p_1 and p_2 represent the probabilities of success in the two experiments. Use a normal approximation and choose $\alpha = .05$.

19 If the percentage of defective parts turned out by a workman during two consecutive weeks was 7 and 9 percent, respectively, and if 500 parts were turned out during each of those weeks, would the inspector be justified in claiming that quality had slipped?

20 Take a sample of 100 random digits from Table II in the appendix by taking the first digit in 100 random numbers listed there, and record the number of 9's obtained. Choosing $\alpha = .10$, use this sample result to test the hypothesis $H_0: p = 1/10$ against $H_1: p \neq 1/10$.

21 Calculate the value of β if $\alpha = .10$, and the best test is employed for testing $H_0: \mu = 10$ against $H_1: \mu = 11$ for a normal variable with $\sigma = 1$, and if a sample of 25 is to be taken.

22 Forty pairs of rats, matched with respect to ability, are given a pill with one member of each pair receiving a stimulant in his pill. Races through a maze are run between each pair. Let X denote the number of races won by the stimulated rats. Construct a best test for testing $H_0: p = 1/2$ against $H_1: p > 1/2$, where p is the probability that a stimulated rat will win a race. Choose a convenient value of α close to .10. Calculate β if $p = .6$ under H_1. Use a normal approximation.

23 Calculate the value of β for testing $H_0: \sigma^2 = 10$ against $H_1: \sigma^2 = 12$ based on a sample of 10 of a normal variable X for which $\mu = 0$ if $\alpha = .10$ and the best critical region is used.

24 If X has the density $f(x \mid \mu) = \dfrac{\exp\left[-\dfrac{1}{2}\left(\dfrac{x-\mu}{10}\right)^2\right]}{10\sqrt{2\pi}}$, how large should n be chosen so that when testing $H_0: \mu = 100$ against $H_1: \mu = 110$ the value of β will be .10 when $\alpha = .05$?

25 Given $f(x \mid \mu) = \dfrac{e^{-\mu}\mu^x}{x!}$, $x = 0, 1, \ldots$, and given $H_0: \mu = 10$ and $H_1: \mu = 9$, how large should n be chosen to guarantee $\alpha = \beta = .10$ if it is assumed that n is sufficiently large to permit a normal approximation for $\sum X_i$?

26 Let $f(x \mid \mu) = \dfrac{\exp\left[-\dfrac{1}{2}\left(\dfrac{x-\mu}{2}\right)^2\right]}{2\sqrt{2\pi}}$, $H_0: \mu = 0$ and $H_1: \mu = 1$, and choose $n = 16$. If $\overline{X} \geq c$ is selected as the critical region, calculate the values of α and β corresponding to the values $c = 0$, $1/2$, 1, ∞, $-\infty$, and graph these pairs of values to show the relation between α and β for this problem. This graph will illustrate the fact that one of these variables is a decreasing function of the other.

27 Given a sample of size 4 of a normal variable for which $\sigma = 1$, and the hypotheses $H_0: \mu = 0$, $H_1: \mu \neq 0$, choose the two $2\frac{1}{2}$ percent tails of the $\overline{X}$ distribution as critical region, and calculate the values of the power function for this test corresponding to $\mu = 0$, $\pm 1/2$, ± 1, and ± 2. Graph the resulting points, and draw a smooth curve through them to obtain a graph of the power function $P(\theta)$.

28 A box contains 10 items. The box of items will be purchased if a sample of 4 shows at most 1 defective item. Find an expression for the power function of this test as a function of the proportion p of defective items in the box.

29 Let X be a binomial variable with $n = 3$. If the critical region for testing $H_0: p = p_0$ consists of (a) $X = 3$, (b) $X = 2$ or 3, calculate the power function of those two tests and graph them. Is one test better than the other?

30 Sketch the power function of the test that uses as critical region those points for which $|X - 4| \geq 2$, where X is the binomial variable based on $n = 8$ and $H_0: p = 1/2$. Choose the values $p = 1/2, 3/4$, and 1, and use symmetry for $p = 1/4$ and 0.

31 Determine whether the test in Exercise 8 is a uniformly most powerful test for the alternative $H_1: \theta < \theta_0$.

32 Determine whether the test in Exercise 9 is a uniformly most powerful test for the alternative $H_1: \theta < \theta_0$.

33 Show that the test in Exercise 27 is not uniformly most powerful by finding a test with the same value of α that has a smaller value of β for some value of μ.

34 Given $f(x \mid \theta) = 1 + \theta^2[x - (1/2)], 0 \le x \le 1, 0 \le \theta \le \sqrt{2}$,
   (a) find a best critical region of size $\alpha = .1$ for testing $H_0: \theta = 0$ against $H_1: \theta \ne 0$, based on a single value of X.
   (b) Is this a uniformly most powerful test?

35 Choose $\alpha = \beta = .2$ and test the hypothesis $H_0: p = 1/2$ against $H_1: p = 2/3$ sequentially by tossing a coin until a decision is reached. Let p denote the probability of a head showing.

36 By adding 10 to each number obtained in Table III in the appendix, obtain sequential samples of a normal variable with mean 10 and $\sigma = 1$. Use these sample values to test $H_0: \mu = 10$ against $H_1: \mu = 10.5$. The sequential test for this problem is given by (18).

37 Choose $\alpha = .1$ and $\beta = .2$, and construct a sequential test for testing $H_0: \sigma = 10$ against $H_1: \sigma = 12$ for a normal variable with zero mean.

38 Choose $\alpha = \beta = .1$ and construct a sequential test for testing $H_0: \theta = 4$ against $H_1: \theta = 3$ if $f(x \mid \theta) = \theta e^{-\theta x}, x > 0$.

39 Construct a sequential test for testing $H_0: \mu = \mu_0$ against $H_1: \mu = \mu_1 > \mu_0$ for X a Poisson variable.

40 Given $\alpha = \beta = .1$, construct a sequential test for testing $H_0: \mu = 12$,
$$\sigma = 4 \text{ against } H_1: \mu = 10, \sigma = 2, \text{ if } f(x \mid \theta) = \frac{\exp\left[-\frac{1}{2}\left(\frac{x - \mu}{\sigma}\right)^2\right]}{\sigma\sqrt{2\pi}}.$$

41 Calculate $E_{1/2}(n)$ for Exercise 35 and compare with the value of n that resulted from your experiment.

42 Calculate $E_4(n)$ for Exercise 38.

43 Calculate $E_{12}(n)$ for Exercise 37.

44 For the problem that produced Table 1 in Section 3.3, calculate the value of n for a fixed size sample test based on $\overline{X}$ that will yield $\alpha = .05$ and $\beta = .10$. Your result should verify that n is approximately 34.

45 Given that X is a normal variable and given the sample values $n = 16, \bar{x} = 40, s = 5$, test the hypothesis that $\mu = 42$.

46 Take a sample of 20 from Table III in the appendix and use it to test the hypothesis $H_0: \mu = .1$ against $H_1: \mu \ne .1$ by means of the t test.

47 Given $n = 10, \bar{x} = 20, s = 5$ for X a normal variable, find a 95 percent confidence interval for μ.

48 Use the data of Exercise 46 to construct a 90 percent confidence interval for μ by means of the t variable.

49 Given the sample values $n_1 = 10$, $\bar{x}_1 = 20$, $s_1 = 6$, $n_2 = 12$, $\bar{x}_2 = 18$, $s_2 = 5$, for two independent normal variables, test the hypothesis $H_0: \mu_1 = \mu_2$.

50 In an industrial experiment a job was performed by 30 workmen using method A and by 40 workmen using method B. The times required to complete the job for the two groups yielded the following sample values:

$$\bar{x} = 53.7, \quad \bar{y} = 55.2, \quad \sum (x_i - \bar{x})^2 = 106, \quad \sum (y_i - \bar{y})^2 = 172.$$

(a) Test the hypotheses $\mu_A = \mu_B$ assuming that the conditions necessary for the application of the t distribution are satisfied here.

(b) Find a 95 percent confidence interval for $\mu_A - \mu_B$.

51 Find a 90 percent confidence interval for $\mu_1 - \mu_2$ in Exercise 49.

52 Given samples of sizes n_1 and n_2, respectively, from two normal variables with zero means and variances σ_1^2 and σ_2^2, respectively, construct a likelihood ratio test for testing $H_0: \sigma_1^2 = \sigma_2^2$. Show that it can be reduced to a test that chooses as critical region $F < F_1$ and $F > F_2$, where F is the variable $F = \dfrac{\chi_1^2/v_1}{\chi_2^2/v_2}$, and χ_1^2 and χ_2^2 are independent chi-square variables with v_1 and v_2 degrees of freedom, respectively. This F variable was derived in Volume I.

53 Use the result in Exercise 52 and Table VII in the appendix to test $H_0: \sigma_1^2 = \sigma_2^2$ for two normal variables with zero means, if samples from them yielded $n_1 = 10$, $\sum x_i^2 = 1440$, $n_2 = 20$, and $\sum y_i^2 = 6480$.

54 The same type of derivation as that in Exercise 52 will show that the only modification needed to test that hypothesis when the two theoretical means are unknown is to choose $v_1 = n_1 - 1$ and $v_2 = n_2 - 1$ for the F variable, which now assumes the form

$$F = \frac{\sum (x_i - \bar{x})^2/(n_1 - 1)}{\sum (y_i - \bar{y})^2/(n_2 - 1)}.$$ Use this result to test $H_0: \sigma_1^2 = \sigma_2^2$ for two normal variables if samples from them yielded $n_1 = 10$, $s_1 = 12$, $n_2 = 20$, and $s_2 = 18$.

55 Show that the variable t^2 with v degrees of freedom is a special case of the F variable with $v_1 = 1$ and $v_2 = v$.

56 Construct a likelihood ratio test for testing $H_0: \mu = \mu_0$ against $H_1: \mu \neq \mu_0$ for a normal variable if σ is known. Compare this test with an earlier test for this problem.

57 Construct a likelihood ratio test for testing $H_0: \sigma = \sigma_0$ against $H_1: \sigma \neq \sigma_0$ for a normal variable if μ is known. Compare this test with an earlier test for this problem.

58 According to Mendelian inheritance, offspring of a certain crossing should be colored red, black, or white in the ratios $9:3:4$. If an experiment produced 150, 70, and 100 offspring in those categories, is the theory justified here?

59 Toss a coin 100 times and apply the chi-square test to see whether the coin is honest.

60 Take a sample of 200 random digits from Table II in the appendix, recording the number of times each digit was obtained. Apply the chi-square test to see whether these frequencies are compatible with the hypothesis $H_0: p_i = 1/10$, $i = 0, \ldots, 9$.

61 In an experiment with flowers of a certain species, theory predicts that there should be four flower types in the ratios $9:3:3:1$. If an experiment produced the frequencies 120, 50, 40, 10 for these four types, is it compatible with the theory?

62 According to the Hardy-Weinberg formula, the number of flies resulting from certain crossings should be in the proportions $q^2 : 2pq : p^2$ where $p + q = 1$. If an experiment yielded the frequencies 80, 90, 40, is it compatible with this formula with $q = .6$?

63 In Exercise 62 estimate q by means of the maximum likelihood formula
$$\hat{q} = \frac{n_1 + (1/2)n_2}{n_1 + n_2 + n_3}$$, where n_1, n_2, and n_3 are the frequencies observed in the three cells. Apply the chi-square test to the data of Exercise 62 with q replaced by $\hat{q}$ to see whether the data are compatible with the theory when q is not specified.

64 Fit a Poisson density to the following famous data on the number of deaths that resulted per year in an army corps for 10 Prussian cavalry army corps over a period of 20 years from being kicked by a horse. The total number of sampling units here is therefore 200. Apply the chi-square test to see whether the Poisson assumption is substantiated.

Number of deaths during the year	0	1	2	3	4
Observed frequency	109	65	22	3	1

65 Fit a binomial density with $n = 4$ to the following data. Apply the chi-square test to see whether the binomial assumption is substantiated.

x	0	1	2	3	4
f	10	40	60	50	16

66 The following data are for a sample of 300 car owners who were classified with respect to age and the number of accidents they had

during the past two years. Test to see whether there is any relationship between these two variables.

Accidents

		0	1 or 2	3 or more
	≤ 21	8	23	14
Age	$22 - 26$	21	42	12
	≥ 27	71	90	19

67 Show that the solution of the equations obtained by setting the partial derivatives of the likelihood function with respect to the variables $p_{i.}$, $i = 1, \ldots, r - 1$, and $p_{.j}$, $j = 1, \ldots, c - 1$, for a contingency table is that given in the text.

68 Given three normal densities with unit variances and with means $-1, 0, 1$, respectively, find the Bayes solution to the multiple decision problem based on a single observed value of X when
(a) $\pi_1 = \pi_2 = \pi_3 = 1/3$,
(b) $\pi_1 = 4/7, \pi_2 = 2/7, \pi_3 = 1/7$.

69 Given three exponential densities $f(x \mid \theta) = \theta e^{-\theta x}$, $x > 0$, with $\theta = 1, 2$, and 3, respectively, find the Bayes solution to the multiple decision problem if $\pi_1 = \pi_2 = \pi_3 = 1/3$ and a single observation is to be taken.

70 Given three normal densities $f(x, y \mid \theta) = \dfrac{1}{2\pi} e^{-(1/2)[(x-\mu)^2 + (y-v)^2]}$ determined by

μ	-1	0	1
v	0	1	0

,

find the Bayes solution to the multiple decision problem if $\pi_1 = \pi_2 = 1/4$ and $\pi_3 = 1/2$ and if a single observation (x, y) is to be taken. Graph your results in the x, y plane.

4 | *Linear Models— Estimation*

Most of the problems of estimation in Chapter 2 were concerned with the estimation of a single parameter of a probability density function. The problem of estimating a vector parameter was solved by resorting to the method of maximum likelihood. This was justified on the grounds that it can be shown that joint maximum likelihood estimates possess certain optimal properties. In this chapter we shall return to the vector parameter estimation problem, but we shall restrict ourselves to what are known as linear model problems. Such problems are also called regression problems, and they constitute some of the most useful applications of statistical theory.

A simple illustration of the type of problem that can be treated under a linear model setup is the problem discussed briefly in Section 2.9 of predicting the height of a son from knowing the height of the father. There the random variables X and Y represented the heights of the father and son, respectively. The best predictor, based on a squared error loss function, of Y for $X = x$ was shown to be given by

$$d(x) = E(Y \mid x).$$

Since $E(Y \mid x)$ is the mean value of the random variable Y corresponding to the fixed value $X = x$, the graph of $d(x)$ as a function of x is a curve representing the locus of such mean points. Experience with data on the relationship between the height of a father (X) and son (Y) indicates that it would be reasonable to assume that $d(x)$ is a linear function of x. Thus, for this problem we shall assume that

(1)
$$d(x) = a + bx,$$

where a and b are unknown parameters whose values must be estimated before the value of Y can be predicted. If we were given the conditional density $f(y \mid x)$ for these two random variables, we could use it to find $E(Y \mid x)$. In that case it would not be necessary to estimate any parameters, even though $E(Y \mid x)$ turned out to be a linear function of x. It is seldom actually true, however, that $f(y \mid x)$ is completely specified; consequently, the problem of estimating unknown parameters is likely to arise in both cases.

Although $d(x)$ in (1) is a linear function of x, it is the fact that it is linear in the unknown parameters a and b that makes this a linear model problem. Thus, the function $d(x) = a + be^x$ is not a linear function of x, but is linear in the parameters a and b, and therefore is a model that can be treated by the methods of this chapter. The function $d(x) = a + be^{-cx}$, which is a very important function arising in the study of biological growth, is an example of a nonlinear model because of the location of the parameter c. The theory of such models is much more difficult than that for linear models, and therefore we restrict ourselves to linear models.

The second problem discussed in Section 2.9 can also be treated under our linear model setup, provided that we are willing to make certain realistic assumptions. The problem is to try to predict the freshman year grade point average of an entering college student on the basis of his high school grade point average, his scholastic aptitude test score, and his score on a placement test given by the college. If Y, X_1, X_2, and X_3 are random variables representing these quantities for an entering freshman, then as before the best predictor of Y for the fixed values $X_1 = x_1$, $X_2 = x_2$, and $X_3 = x_3$ is given by $d(x) = E(Y \mid x_1, x_2, x_3)$. If we assume that $d(x)$ is a linear function of its arguments, and experience with college records indicates that this is a reasonable assumption, then our problem is one of estimating the unknown parameters in

$$d(x) = a_1x_1 + a_2x_2 + a_3x_3 + a_4.$$

As before, even if we were given the information that $E(Y \mid x_1, x_2, x_3)$ is a linear function of the x's, it is usually unrealistic in a practical situation to assume that we also know the values of the parameters of $d(x)$; consequently, the problem of estimating the a's is likely to arise in this situation as well.

As a function of the variables x_1, x_2, and x_3, the function $d(x)$ is commonly called a *linear regression function*. From our point of view, however, the important feature of this model is that it is linear in the parameters a_1, a_2, a_3, and a_4, and therefore is the type of model that we wish to study in this chapter.

4.1. Simple linear regression

In the discussion of the problem of predicting a son's height from the father's height, it was assumed that both X and Y were random variables and that Y was to be predicted on the basis of being told the value of X. Our point of view in this section is somewhat different. We do not necessarily treat X as a random variable, but instead assign it values at pleasure and then attempt to predict the value of Y corresponding to any assigned value of X. For example, we might select a set of heights covering the normal range of male heights and then proceed to choose one individual at random from each group of individuals having one of the selected father heights. This approach has the practical advantage of enabling the experimenter to concentrate or restrict his sampling to those individuals whose X values interest him, rather than taking random pairs X, Y. It has the additional advantage of making it unnecessary to know the distribution of X in carrying out theoretical investigations. Although we are proposing to choose our X values in advance of sampling the Y's, we may, and often do, choose random pairs (X, Y) and treat the resulting X's as the assigned values.

As an illustration of how to proceed in the estimation of a and b in (1), suppose that we wish to predict the value of a freshman's grade point average on the basis of his high school grade point average only, and that the regression function is assumed to be linear. For this purpose we shall select a set of high school grade point averages $x_1, \ldots, x_n$ that are uniformly spaced over the range of values of high school averages. Then corresponding to each x_i value, $i = 1, \ldots, n$, we shall choose one student at random from those in the sophomore class who had x_i for their high school grade point average. Let y_i denote this student's freshman year grade point average. This set of values (x_i, y_i), $i = 1, \ldots, n$, may then be used to estimate the parameters a and b in (1).

Our first estimation procedure will be based on assuming that the values $y_1, \ldots, y_n$ are the observed values of a set of independent random variables $Y_1, \ldots, Y_n$, each of which is normally distributed with the same variance σ^2 and with means given by $E(Y_i \mid x_i) = a + bx_i$, $i = 1, \ldots, n$. As a result the joint density of these variables is given by

$$
(2) \qquad \prod_{i=1}^{n} f(y_i \mid x_i) = \prod_{i=1}^{n} \frac{\exp\left[-\dfrac{1}{2\sigma^2}(y_i - a - bx_i)^2\right]}{\sqrt{2\pi}\sigma}
$$

$$
= \frac{\exp\left[-\dfrac{1}{2\sigma^2}\sum_{i=1}^{n}(y_i - a - bx_i)^2\right]}{(2\pi)^{n/2}\sigma^n}.
$$

We shall estimate the parameters a and b by the method of maximum likelihood. Taking logarithms will give

$$
\log \prod_{i=1}^{n} f(y_i \mid x_i) = -\log (2\pi)^{n/2}\sigma^n - \frac{1}{2\sigma^2}\sum_{i=1}^{n}(y_i - a - bx_i)^2.
$$

Differentiating with respect to a and b, setting those derivatives equal to zero, and simplifying, we will obtain the equations

$$
\sum_{i=1}^{n}(y_i - a - bx_i) = 0
$$

$$
\sum_{i=1}^{n}(y_i - a - bx_i)x_i = 0.
$$

These equations are equivalent to the equations

$$(3) \qquad an + b \sum_{i=1}^{n} x_i = \sum_{i=1}^{n} y_i$$

$$a \sum_{i=1}^{n} x_i + b \sum_{i=1}^{n} x_i^2 = \sum_{i=1}^{n} x_i y_i.$$

It can be shown by calculus techniques, or otherwise, that the unique solution of these equations, if it exists, yields an absolute maximum as well as a calculus relative maximum of the likelihood function (2). This will be demonstrated in a later section. The solution of these equations when substituted into (1) will then yield the desired estimate of our linear regression function. The function can then be used to predict the grade point average of any entering freshman.

Suppose that we have no knowledge about the nature of $f(y \mid x)$ and are unwilling to assume that it is a normal density. If we are still happy with the assumption that $E(Y \mid x)$ is a linear function of x and that mean squared error is a valid basis for good estimation, we can then find an empirical solution to the problem by choosing a and b to minimize the empirical mean squared error of the sample. For a fixed a and b, the quantity $a + bx_i$ is the predicted value of Y_i, and y_i is its observed value; therefore the error of prediction is $y_i - a - bx_i$, and the empirical mean squared error is given by

$$\frac{1}{n} \sum_{i=1}^{n} (y_i - a - bx_i)^2.$$

We therefore choose a and b to minimize this quantity. Differentiating with respect to a and b will yield the same equations as those in (3), which were obtained by maximum likelihood methods based on a normality assumption. Equations (3) are called the *normal equations of least squares*, and the technique of estimation employed here is called the *method of least squares* for simple linear regression. It is a strictly empirical approach to estimation that has been popular for over a century.

From a geometrical point of view, the method of least squares as it is applied to this problem is a method of fitting a straight line $y = a + bx$ to a set of points (x_i, y_i), $i = 1, \ldots, n$, in the x, y plane. The method consists in choosing the straight line that minimizes the sum of the squares of the vertical distances of the points from the line.

As a numerical illustration, consider the problem of fitting a straight line to the set of points corresponding to the data of Table 1. These data represent the scores made on an aptitude test (x) and an achievement test (y) for a group of thirteen students selected on the basis of their aptitude test scores. These pairs of numbers have been plotted as points in the x, y plane of Figure 1.

Table 1

x	70	75	80	85	90	95	100	105	110	115	120	125	130
y	30	26	51	48	40	46	61	76	61	50	64	53	71

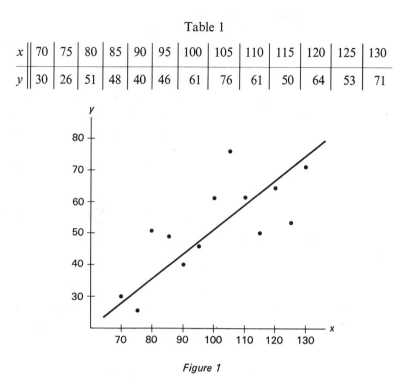

Figure 1

The formulas of least squares become somewhat simpler if the variable x is measured from its empirical mean $\bar{x}$. This means that we write the equation of a straight line in the form $y = a + b(x - \bar{x})$. The normal equations (3) then assume the form

$$an + b \sum (x_i - \bar{x}) = \sum y_i$$
$$a \sum (x_i - \bar{x}) + b \sum (x_i - \bar{x})^2 = \sum (x_i - \bar{x}) y_i.$$

But $\sum (x_i - \bar{x}) = 0$; consequently, the least squares estimates of a and b for this version of the straight line are given by

(4) $$\hat{a} = \bar{y} \quad \text{and} \quad \hat{b} = \frac{\sum (x_i - \bar{x}) y_i}{\sum (x_i - \bar{x})^2}.$$

Recalling the formula in Section 2.6 that defines the sample correlation coefficient r, it will be observed that

$$\hat{b} = r \frac{s_y}{s_x}.$$

As a result, the equation of the least squares line can be written in the form

$$y = \bar{y} + r \frac{s_y}{s_x} (x - \bar{x}).$$

If we compare this result with the equation of the best linear predictor $d(x)$ given by Formula (31) in Section 2.9, we see that our least squares predictor is the sample estimate of $d(x)$ obtained by replacing μ_x, μ_y, σ_x, σ_y, and ρ in that formula by their sample estimates $\bar{x}$, $\bar{y}$, s_x, s_y, and r.

The computations needed to apply our formula to the data of Table 1 are most easily carried out after $\bar{x} = 100$ is subtracted from each x entry:

$x_i - \bar{x}$	-30	-25	-20	-15	-10	-5	0	5	10	15	20	25	30
y_i	30	26	51	48	40	46	61	76	61	50	64	53	71

Calculations will produce the values

$$\bar{y} = 52.1, \qquad \sum (x_i - \bar{x})y_i = 2{,}555, \qquad \sum (x_i - \bar{x})^2 = 4{,}550.$$

These in turn will yield the least squares estimates

$$\hat{a} = 52.1 \quad \text{and} \quad \hat{b} = .56.$$

The desired estimate of the linear predictor for predicting a student's achievement on the basis of his aptitude test score is therefore given by the linear function

$$y = 52.1 + .56(x - 100).$$

Or, if form (1) is preferred, the equation becomes

$$y = -3.9 + .56x.$$

A graph of this least squares line is shown in Figure 1.

A determination of the reliability of an empirical predictor such as this requires a knowledge of the accuracy of $\hat{a}$ and $\hat{b}$ as estimators of a and b, and also a knowledge of σ^2, which gives the mean squared error of prediction. The accuracy of $\hat{a}$ and $\hat{b}$ will be studied in Section 4.5; therefore we shall be concerned here only with how to obtain information about σ^2 when its value is not known.

If we assume the normal model given by (2), we can obtain the maximum likelihood estimate of σ^2. Taking the logarithm of (2), differentiating with respect to σ, and setting this derivative equal to zero will yield the estimate

$$\hat{\sigma}^2 = \frac{1}{n} \sum_{i=1}^{n} (y_i - \hat{a} - \hat{b}x_i)^2,$$

where $\hat{a}$ and $\hat{b}$ denote the estimates determined by equations (3). If the modified model with x_i replaced by $x_i - \bar{x}$ is used, then $\hat{a}$ and $\hat{b}$ will represent the estimates given by (4). We may use this formula to estimate σ^2 even without the normality assumption, provided we assume that the Y's possess a common variance σ^2. We would then be using the method of moments to estimate σ^2.

Under the normality assumption (2), we can make probability statements concerning the magnitude of the prediction error corresponding to any of the selected x values. For example, we could make the probability statement $P(a + bx_i - 2\sigma < Y_i < a + bx_i + 2\sigma) = .95$ for any i, which means that the probability is .95 that the prediction error $|Y_i - a - bx_i|$ will not exceed 2σ. If we replace the parameters a, b, and σ by their sample estimates, we will only be able to approximate such probabilities. The methods that will be presented in the next chapter, however, are capable of solving such problems with exact probabilities.

For the preceding illustration, calculations based on the empirical predictor $y = -3.9 + .56x$ will yield the predicted values, denoted by y_i', and the differences between the observed and predicted values, denoted by e_i, that are shown in Table 2.

Table 2

y_i	30	26	51	48	40	46	61
v_i'	35.3	38.1	40.9	43.7	46.5	49.3	52.1
e_i	-5.3	-12.1	10.1	4.3	-6.5	-3.3	8.9

76	61	50	64	53	71
54.9	57.7	60.5	63.3	66.1	68.9
21.1	3.3	-10.5	.7	-13.1	2.1

Calculations based on these values will yield the estimate $\hat\sigma = 9.5$. It will be observed, for example, that 5 of the 13 differences in Table 2 exceeded $\hat\sigma = 9.5$. This corresponds to 38 percent, which differs little from the 32 percent to be expected for a normal model when the parameters a, b, and σ are known.

4.2. Simple nonlinear regression

In the preceding section we assumed that $E(Y \mid x)$ was a linear function of x. Now consider the problem when this assumption is not justified. If we are given the conditional density $f(y \mid x)$, we can calculate $E(Y \mid x)$ and determine what type of function of x it is. For this more general situation the function $d(x) = E(Y \mid x)$ is called the *regression function of Y on x*, and its graph is called the *regression curve*. Since $E(Y \mid x)$ is the mean value of the random variable Y corresponding to the fixed value x, the regression curve is the locus of the means of the conditional densities $f(y \mid x)$ as x varies over its domain of values.

As an illustration of nonlinear regression, suppose the random variable Y possesses the conditional density

$$f(y \mid x) = \frac{\exp\left[-(1/2)(y - ax^2)^2\right]}{\sqrt{2\pi}}.$$

Since this is a normal density with mean ax^2, it follows that $d(x) = E(Y \mid x) = ax^2$. Hence, the regression curve is the parabola $d(x) = ax^2$, and the conditional random variable Y corresponding to any assigned value of x possesses a normal distribution with unit variance and with its mean lying on this parabola. These facts are shown geometrically in Figure 2.

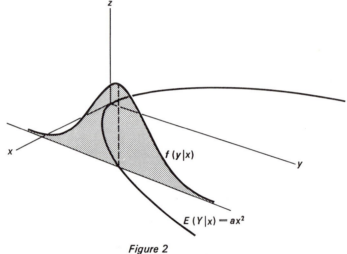

Figure 2

If the value of the parameter a were unknown, it could be estimated by the method of maximum likelihood or the method of least squares. Because of the normality assumption, these two methods would yield the same estimate. Incidentally, although this is a nonlinear regression function, it still constitutes a linear model because of the nature of the parameter a.

To observe the difficulties that arise when we do not restrict ourselves to linear models, consider the problem of estimating the regression function $E(Y \mid x) = a + be^{-cx}$ on the basis of a set of observed values $\{x_i, y_i\}$, $i = 1, \ldots, n$. If we use the method of least squares, it is necessary to find the values of a, b, and c that minimize the function

$$G(a, b, c) = \sum_{i=1}^{n} [y_i - a - be^{-cx_i}]^2.$$

Differentiating G with respect to a, b, and c, and setting those derivatives equal to zero will, after a few algebraic simplifications, yield the equations

$$\sum [y_i - a - be^{-cx_i}] = 0$$

$$\sum [y_i - a - be^{-cx_i}]e^{-cx_i} = 0$$

$$\sum [y_i - a - be^{-cx_i}]x_ie^{-cx_i} = 0.$$

These in turn are equivalent to the equations

$$an + b \sum e^{-cx_i} = \sum y_i$$

$$a \sum e^{-cx_i} + b \sum e^{-2cx_i} = \sum e^{-cx_i}y_i$$

$$a \sum x_ie^{-cx_i} + b \sum x_ie^{-2cx_i} = \sum x_ie^{-cx_i}y_i.$$

The difficulty in solving these equations rests in the nonlinear position of c. Since there is no direct way of solving these equations, it is necessary to resort to iterative numerical methods. Thus, if an approximate solution a_0, b_0, c_0 could be obtained by some method, say by trial and error, it would be possible to obtain an improved approximation by writing $a = a_0 + \alpha$, $b = b_0 + \beta$, $c = c_0 + \gamma$, substituting these into the equations, approximating $e^{-\gamma x_i}$ by $1 - \gamma x_i$, and then solving the resulting linear equations for α, β, and γ. This iterative technique could be repeated until a stable solution is obtained. This problem illustrates very nicely the difficulties that arise in finding estimates for nonlinear models. If modern computing facilities are available, such numerical difficulties are not serious; however, the difficulty of the theory is quite another matter.

The method of least squares is a very general method for estimating parameters of any type of regression function. For the particular problems that we considered, least squares and maximum likelihood produced the same estimates; however this is not true in general. The method of least squares is also the fundamental basis for approximating functions by simpler functions. The student of statistics should not entertain the mistaken notion that it is merely an empirical device that is introduced for estimation when the method of maximum likelihood is not available. It is far more important and useful than this.

4.3. Multiple linear regression

The two methods of estimation, namely maximum likelihood and least squares, which were discussed in the preceding sections, can be applied in the same manner to more general linear models. For example, the problem of predicting a freshman student's grade point average on the basis of his high school record, his college aptitude test score, and his placement test score is easily solved by the preceding methods if it is assumed that

(5) $$E(Y \mid x_1, x_2, x_3) = a_1x_1 + a_2x_2 + a_3x_3 + a_4,$$

where these variables are the ones defined in the introduction to this chapter. In that earlier discussion the x's were assumed to be the observed values of corresponding random variables; however, in linear model theory the x's may be assigned values at pleasure, and then the only random variable is Y. Thus, a set of values (x_{i1}, x_{i2}, x_{i3}), $i = 1, \ldots, n$, may be selected in advance, and then a student selected at random from each of the n groups of students who possess one of those sets of x values. If y_i denotes the value of Y for such a randomly selected student from the ith group, then the set of values $(y_i, x_{i1}, x_{i2}, x_{i3})$, $i = 1, \ldots, n$, can be used for obtaining estimates of the a's.

Although the x's are not being treated here as random variables, it would be permissible to select n students at random from the student body and use the resulting set of values for obtaining the desired estimates. However, in attempting to determine the accuracy of our estimates, it is necessary to assume that in each of the repeated sampling experiments the same set of x values will be employed. This follows because only the conditional density $f(y \mid x_1, x_2, x_3)$ is used to determine the accuracy of the estimates. For this density the values of x_1, x_2, and x_3 are fixed, and Y is the random variable. But since the x's can be given any desired values, those values can be assigned in advance of the experiment, or they can be obtained from an unrestricted random sample of n students.

For the purpose of continuing the discussion of how to estimate the parameters in (5), let $(y_i, x_{i1}, x_{i2}, x_{i3})$, $i = 1, \ldots, n$, denote the values that were obtained from a random sample of n students. We shall assume as before that the random variables $Y_1, \ldots, Y_n$ are independently normally distributed with a common variance σ^2 and with means given by

$$E(Y_i \mid x_{i1}, x_{i2}, x_{i3}) = a_1 x_{i1} + a_2 x_{i2} + a_3 x_{i3} + a_4.$$

As a consequence, the joint density of these variables is given by

(6) $\displaystyle\prod_{i=1}^{n} f(y_i \mid x_{i1}, x_{i2}, x_{i3})$

$$= \prod_{i=1}^{n} \frac{\exp\left[-\dfrac{1}{2\sigma^2}(y_i - a_1 x_{i1} - a_2 x_{i2} - a_3 x_{i3} - a_4)^2\right]}{\sqrt{2\pi}\,\sigma}$$

$$= \frac{\exp\left[-\dfrac{1}{2\sigma^2}\displaystyle\sum_{i=1}^{n}(y_i - a_1 x_{i1} - a_2 x_{i2} - a_3 x_{i3} - a_4)^2\right]}{(2\pi)^{n/2}\sigma^n}.$$

The maximum likelihood estimates of the a's are readily obtained by taking logarithms, differentiating with respect to each of the a's, setting those derivatives equal to zero, and solving the resulting equations. The equations obtained in this manner, after some algebraic simplifications, are the following

(7)
$$\sum_{i=1}^{n} (y_i - a_1 x_{i1} - a_2 x_{i2} - a_3 x_{i3} - a_4) = 0$$

$$\sum_{i=1}^{n} (y_i - a_1 x_{i1} - a_2 x_{i2} - a_3 x_{i3} - a_4) x_{i1} = 0$$

$$\sum_{i=1}^{n} (y_i - a_1 x_{i1} - a_2 x_{i2} - a_3 x_{i3} - a_4) x_{i2} = 0$$

$$\sum_{i=1}^{n} (y_i - a_1 x_{i1} - a_2 x_{i2} - a_3 x_{i3} - a_4) x_{i3} = 0.$$

If no distribution assumptions are made on the Y's and the empirical least squares approach is used, precisely the same equations will be obtained, because the sum of squares of the prediction errors is the sum of squares occurring in the exponent of (6). As before, this equivalence occurs because of our normality assumption on the Y's.

In view of the equivalence of the maximum likelihood estimates and the least squares estimates here, and the fact that the latter require fewer assumptions, the natural question to ask is "Why not be satisfied with least squares estimates?" If we are satisfied with point estimates of the coefficients and do not wish to discuss their accuracy or make any probability statements about the errors of prediction, then least squares techniques will suffice; otherwise it is necessary to introduce some probability distribution assumption about the random variables $Y_1, \ldots, Y_n$. Furthermore, if we wish to test hypotheses about some of the parameters, and we shall be doing that in the next chapter, then we also need to make some distribution assumption about the Y's.

As in the case of a single variable, the least squares equations given by (7) are simpler to solve if the x's are measured from their empirical means, because then the number of nontrivial equations is reduced by one. The fact that a solution of equations (7), if it exists, actually minimizes the sum of squares will be demonstrated in the next section.

The following material requires some knowledge of matrix methods. A review of the parts of matrix theory that are needed in this and the next chapter is available in the appendix.

4.4. Matrix methods

The general linear regression model of which (5) is a special case can be written in very compact form by using vector and matrix notation. Toward this end, we shall first write the regression values in the form

(8)
$$E(Y_i) = \beta_1 x_{i1} + \cdots + \beta_k x_{ik}, \qquad i = 1, \ldots, n.$$

This is a linear regression function with k fixed variables. A constant term, such as a_4 in (5), can be obtained here by choosing $x_{ik} = 1$,

$i = 1, \ldots, n$, so this form is completely general. We now introduce the following column vectors and matrix, where the symbol X should not be confused with the random variable X of the introduction to this chapter. The following notation involving X is traditional in statistical theory. Since only the Y's are random here, there should be no confusion with our earlier use of X to represent a random variable.

$$Y = \begin{bmatrix} Y_1 \\ \vdots \\ Y_n \end{bmatrix}, \qquad \beta = \begin{bmatrix} \beta_1 \\ \vdots \\ \beta_k \end{bmatrix}, \qquad X = \begin{bmatrix} x_{11} \cdots x_{1k} \\ \vdots \qquad \vdots \\ x_{n1} \cdots x_{nk} \end{bmatrix}.$$

In terms of this notation, (8) can be expressed in the compact matrix form

$$(9) \qquad\qquad E(Y) = X\beta.$$

The sum of the squares of the errors of prediction needed for least squares estimation is

$$\sum_{i=1}^{n} (y_i - \beta_1 x_{i1} - \cdots - \beta_k x_{ik})^2.$$

In terms of our matrix notation this becomes

$$(y - X\beta)'(y - X\beta),$$

where y denotes the sample value of the random vector Y, and where a prime on a vector or matrix denotes its transpose. The equations of least squares analogous to (7) are then given by

$$(10) \qquad\qquad (y - X\beta)'X = 0.$$

These equations can be solved in matrix form by multiplying out the parentheses and solving for the vector β. Thus,

$$y'X - \beta'X'X = 0,$$

or

$$\beta'X'X = y'X.$$

Taking the transpose of both sides will give

$$X'X\beta = X'y.$$

Under the assumption that the matrix $X'X$ is nonsingular, the desired solution is given by the formula

$$(11) \qquad\qquad \hat{\beta} = (X'X)^{-1}X'y.$$

Here $\hat{\beta}$ is a numerical vector, but when y is replaced by the random vector Y, this formula defines the random variable $\hat{\beta}$.

As an illustration of the use of this formula, consider the problem of finding the least squares estimates of the parameters for the regression function $E(Y) = a_1x_1 + a_2x_2 + a_3$, where the variables Y, x_1, and x_2 represent, respectively, car mileage, the units of ingredient A added to the gasoline, and the units of ingredient B added. A controlled experiment, which assigned 0, 1, and 2 units of the two ingredients in all possible combinations for testing, produced the following data expressed in terms of convenient units.

i	1	2	3	4	5	6	7	8	9
y_i	20.4	19.0	18.9	21.0	22.3	18.7	22.0	24.5	22.4
x_{i1}	0	1	2	0	1	2	0	1	2
x_{i2}	0	0	0	1	1	1	2	2	2

First, we rewrite our regression equation in the form of (8) by choosing $k = 3$ and $x_{i3} = 1$, so that $\beta_3 = a_3$ becomes the constant term. Then we calculate the matrix $X'X$ where

$$X' = \begin{bmatrix} 0 & 1 & 2 & 0 & 1 & 2 & 0 & 1 & 2 \\ 0 & 0 & 0 & 1 & 1 & 1 & 2 & 2 & 2 \\ 1 & 1 & 1 & 1 & 1 & 1 & 1 & 1 & 1 \end{bmatrix}.$$

The result of such calculations is the matrix

$$X'X = \begin{bmatrix} 15 & 9 & 9 \\ 9 & 15 & 9 \\ 9 & 9 & 9 \end{bmatrix}.$$

The inverse is calculated next and turns out to be

(12)
$$(X'X)^{-1} = \begin{bmatrix} 1/6 & 0 & -1/6 \\ 0 & 1/6 & -1/6 \\ -1/6 & -1/6 & 4/9 \end{bmatrix}.$$

Then $X'y$ is calculated to give

$$X'y = \begin{bmatrix} 185.8 \\ 199.8 \\ 189.2 \end{bmatrix}.$$

Finally, we calculate $(X'X)^{-1}X'y$ to obtain

$$\hat{\beta} = \begin{bmatrix} -.57 \\ 1.77 \\ 19.82 \end{bmatrix}.$$

The least squares estimate of the regression function is therefore given by the equation

(13) $y = -.57x_1 + 1.77x_2 + 19.82.$

It would appear from this result that ingredient A is detrimental and ingredient B is beneficial to car mileage. On the average, if the amount of A is held fixed, increasing the amount of B increases mileage; whereas if the amount of B is held fixed, increasing the amount of A decreases mileage.

4.5. Properties of least squares estimators

An advantage of matrix methods is that properties of least squares estimators are easily demonstrated by means of them. These properties will be expressed in the form of several theorems.

Theorem 1 *Any solution of the least squares equations minimizes* $(y - X\beta)'(y - X\beta).$

Proof. To prove this theorem, the sum of squares will be written in the following form, where $\hat{\beta}$ denotes a solution of equations (10).

$$(y - X\beta)'(y - X\beta) = ((y - X\hat{\beta}) + X(\hat{\beta} - \beta))'((y - X\hat{\beta}) + X(\hat{\beta} - \beta))$$
$$= (y - X\hat{\beta})'(y - X\hat{\beta}) + (\hat{\beta} - \beta)'X'(y - X\hat{\beta})$$
$$+ (y - X\hat{\beta})'X(\hat{\beta} - \beta) + (\hat{\beta} - \beta)'X'X(\hat{\beta} - \beta).$$

Since $\hat{\beta}$ is a solution of equations (10), it follows that $(y - X\hat{\beta})'X$ and its transpose $X'(y - X\hat{\beta})$ are zero row and column vectors, respectively, and therefore that the second and third terms on the right must vanish. By letting $Z = X(\hat{\beta} - \beta)$, it will be seen that the fourth term is of the form $Z'Z = \sum z_i^2$, which is nonnegative; consequently

$$(y - X\beta)'(y - X\beta) \geq (y - X\hat{\beta})'(y - X\hat{\beta}).$$

Therefore $\hat{\beta}$ does minimize the sum of squares.   ∎

Notice that no calculus techniques were needed to prove that any solution of the equations obtained by setting the partial derivatives equal to zero produces a true minimum. If $X'X$ is nonsingular, equations (10) will possess a unique solution given by Formula (11). This is the usual situation and the one that we will assume hereafter. However, if $X'X$ is singular and equations (10) possess more than one solution, then any such solution will do.

Theorem 2 *The estimator given by means of* (11) *is an unbiased estimator of* β.

Proof. We use (11) to obtain

$$E(\hat{\beta}) = E(X'X)^{-1}X'Y = (X'X)^{-1}X'EY.$$

The last equality follows from the fact that $\hat{\beta}$, as an estimator of β, is a linear combination of the random variables $Y_1, \ldots, Y_n$, and its expected value is therefore the same linear combination of the expected values of those random variables. From (9) it then follows that

$$E\hat{\beta} = (X'X)^{-1}X'X\beta = \beta. \qquad \blacksquare$$

Theorem 3 *If $Y_1, \ldots, Y_n$ are uncorrelated and possess a common variance σ^2, the covariance matrix of the $\hat{\beta}$'s is given by the formula*

$$E(\hat{\beta} - \beta)(\hat{\beta} - \beta)' = \sigma^2(X'X)^{-1}.$$

Proof. Recall from Volume I that

$$\sigma_{ij} = \text{Cov}(Y_i, Y_j) = E(Y_i - \mu_i)(Y_j - \mu_j), \qquad \text{where} \qquad \mu_i = E(Y_i).$$

If the expected value of a matrix of random variables is defined to be the matrix of the expected values of its elements, then

$$E(Y - \mu)(Y - \mu)' = E \begin{bmatrix} Y_1 - \mu_1 \\ \vdots \\ Y_n - \mu_n \end{bmatrix} [Y_1 - \mu_1, \ldots, Y_n - \mu_n]$$

$$= E \begin{bmatrix} (Y_1 - \mu_1)(Y_1 - \mu_1) \cdots (Y_1 - \mu_1)(Y_n - \mu_n) \\ \vdots \qquad\qquad \vdots \\ (Y_n - \mu_n)(Y_1 - \mu_1) \cdots (Y_n - \mu_n)(Y_n - \mu_n) \end{bmatrix}$$

$$= \begin{bmatrix} \sigma_{11} \cdots \sigma_{1n} \\ \vdots \qquad \vdots \\ \sigma_{n1} \cdots \sigma_{nn} \end{bmatrix}.$$

Thus, the matrix of variances and covariances, which is called the covariance matrix, is given by the formula

$$(14) \qquad\qquad [\sigma_{ij}] = E(Y - \mu)(Y - \mu)'.$$

In particular, if the Y's are uncorrelated with a common variance σ^2, then $\sigma_{ij} = 0$, $i \neq j$, and $\sigma_{ii} = \sigma^2$.

This formula will now be used to calculate the covariance matrix of the $\hat{\beta}$'s. Since $E\hat{\beta} = \beta$, we need

$$\hat{\beta} - \beta = (X'X)^{-1}X'Y - (X'X)^{-1}X'X\beta$$

$$= (X'X)^{-1}X'(Y - X\beta).$$

Therefore, because the matrix $(X'X)^{-1}$ is symmetric,

$$E(\hat{\beta} - \beta)(\hat{\beta} - \beta)' = (X'X)^{-1}X'E(Y - X\beta)(Y - X\beta)'X(X'X)^{-1}.$$

Since it is assumed that the Y's are uncorrelated and possess a common variance σ^2, it follows from (14) that

$$E(Y - X\beta)(Y - X\beta)' = \sigma^2 I.$$

Hence, we obtain the desired result, namely

(15) $E(\hat{\beta} - \beta)(\hat{\beta} - \beta)' = \sigma^2(X'X)^{-1}X'X(X'X)^{-1}$

$$= \sigma^2(X'X)^{-1}. \qquad \blacksquare$$

By means of this formula we can read off the variances and covariances of the $\hat{\beta}$'s, once the inverse of the matrix $X'X$ has been computed.

The accuracy of $\hat{\beta}$ as an estimate of β can be determined by means of this covariance matrix, provided that we know the value of σ and provided that we are willing to assume that the Y's are independent normal variables. As an illustration of this use, consider the earlier problem that resulted in the estimated regression equation (13) and the inverse matrix (12). For that problem it follows directly from (12) and (15) that

$$V(\hat{\beta}_1) = \frac{\sigma^2}{6}, \qquad V(\hat{\beta}_2) = \frac{\sigma^2}{6}, \qquad V(\hat{\beta}_3) = (4/9)\sigma^2.$$

From (11) it will be seen that the $\hat{\beta}$'s are normal variables because they are linear combinations of the Y's, which are assumed here to be independent normal variables. A 95 percent confidence interval for the coefficient β_1, for example, would therefore be given by the two numbers

$$\hat{\beta}_1 \pm 1.96\sigma_{\hat{\beta}_1} = -.57 \pm 1.96 \frac{\sigma}{\sqrt{6}}.$$

If σ were unknown here, an approximation to this interval could be obtained by replacing σ by its maximum likelihood estimate $\hat{\sigma}$, where $\hat{\sigma}^2 = (y - X\hat{\beta})'(y - X\hat{\beta})/n$.

Another attractive property of least squares estimators, but which will not be demonstrated here, is that they possess minimum variance among all linear unbiased estimators. This means that $V(\hat{\beta}_i) \leq V(\beta_i^*)$, $i = 1, \ldots, k$, where β_i^* is any linear unbiased estimator of β_i. In view of our use in Chapter 2 of the variance as a basis for good estimation, this property is a very desirable one.

4.6. Some special models

The linear regression model (8) is very versatile because it includes as special cases many important models not customarily associated with the name linear. Thus, if we choose the variable x_j to be the variable given by $x_j = t^{j-1}$, our regression function will assume the form

$$E(Y_i) = \beta_1 + \beta_2 t_i + \cdots + \beta_k t_i^{k-1},$$

which is a polynomial of degree $k - 1$ in t_i. The problem of estimating the β's then reduces to the problem of fitting a polynomial of degree $k - 1$ to a set of n points in the (t, y) plane by the method of least squares. The solution to the problem is given by (11) with $x_{ij} = t_i^{j-1}$.

Another special case of much importance is that of the partial sum of a Fourier series given by

$$a_0 + a_1 \cos t + \cdots + a_r \cos rt$$

$$+ b_1 \sin t + \cdots + b_r \sin rt.$$

This function can be expressed in the form of (8) by choosing $k = 2r + 1$, $x_1 = 1$, $x_j = \cos(j - 1)t$, $j = 2, \ldots, r + 1$, and $x_j = \sin(j - r - 1)t$, for $j = r + 2, \ldots, 2r + 1$. As before, the solution to the problem of fitting a function of this type to a set of points in the (t, y) plane is given by Formula (11). A regression function of this kind is appropriate when dealing with periodic functions and with functions that are obtained as solutions of certain differential equations of mathematical physics and engineering.

Not only does the linear regression model (8) enable us to treat non-linear problems of a single variable, which has been denoted here by t, but it can be modified to include many nonlinear problems in several variables. For example, in our earlier problem of attempting to predict a student's college success by means of three variables x_1, x_2, and x_3, we assumed that a linear function of those variables represented the relationship realistically; however, if this is not the case, we could introduce quadratic terms in those variables to produce a more flexible function. Thus, we might choose the function

$$E(Y) = a_1 x_1 + a_2 x_2 + a_3 x_3 + a_4 + a_5 x_1^2 + a_6 x_2^2 + a_7 x_3^2$$

$$+ a_8 x_1 x_2 + a_9 x_1 x_3 + a_{10} x_2 x_3.$$

In spite of its nonlinear character, this function can also be treated as a special case of model (8) by letting $k = 10$ and choosing $x_4 = 1$, $x_5 = x_1^2$, $x_6 = x_2^2$, $x_7 = x_3^2$, $x_8 = x_1 x_2$, $x_9 = x_1 x_3$, and $x_{10} = x_2 x_3$. Since there is no restriction on the nature of the variables in (8), such choices are possible, and the usual least squares formulas corresponding to model (8) are applicable. Applications such as this enable us to determine whether a simple linear regression function will suffice for making predictions or whether a more sophisticated model is needed. This problem will be discussed further in the next chapter.

4.7. Analysis of variance model

The regression problems that have been considered thus far have been problems in which the variables $x_1, \ldots, x_k$ could be assigned any desired values in certain natural intervals of values. There are many important experiments, however, in which these variables represent only a classification into a few groups. Thus, in attempting to predict a college student's grade point average, we might wish to include such qualitative variables as whether the student attended a private or public high school, or whether his high school was a small, medium, or large size school. As another illustration, in an industrial experiment we might wish to study the effect on productivity of five different types of machines that can be used at one stage of the production process. Here the variable representing machine type is certainly not the kind of variable that can be assigned values at pleasure. We may not even be able to arrange the machine types in a natural order.

For problems of the preceding type in which the variables associated with the random variable Y are in classes or categories rather than being measured, there is a useful regression model called the *linear analysis of variance model*. The technique employed in this model is to analyze the sample variance of the random variable Y into components that can be attributed to the various controlled variables involved in the experiment. Such components are used to determine whether there are meaningful differences in the various categories of a qualitative control variable.

As an illustration, suppose that a firm wishes to buy a set of calculating machines, but is undecided as to which of four certain brands is best suited to its particular type of work, or even whether there is any essential difference in the four brands. It could set up an experiment by assigning several of its employees a set of problems to be solved by the various machines and recording the total times required. Here there are two important variables that can be controlled, namely the brand of machine and the operator. The first problem to be solved then is to test whether the brand differences in total times are meaningful. If the test decides that they are, the next problem is to estimate those differences and to determine the accuracy of the estimates. The solution of those two problems requires an analysis of the variability of the basic random variable, namely total time, into components that can be attributed to brand differences, to operator differences, and to all other factors. The problem of deciding whether brand differences are sufficiently large to be meaningful will be studied in the next chapter on testing; here we shall consider only the problem of estimating such differences.

In order to analyze an experiment of the preceding type in which there are two basic variables to be controlled, it is convenient to represent the experimental results in a two-way table. In such a table there will be four columns to represent the four brands of machines, and there will be as many rows as there are operators assigned to work the problems. In this type of experiment it is assumed that each operator works all problems on each of the four brands of machines. To eliminate the possible advantage of learning a problem through repetition, the order in which the problems were presented to the various operators could be randomized and they could be modified slightly, or a time delay could be introduced between solutions of the same problem. The general setup for a problem of this type is shown in Table 2.

Table 2

$$
\begin{array}{ccc}
y_{11} \cdots y_{1j} \cdots y_{1c} \\
\vdots \qquad \vdots \qquad \vdots \\
y_{i1} \cdots y_{ij} \cdots y_{ic} \\
\vdots \qquad \vdots \qquad \vdots \\
y_{r1} \cdots y_{rj} \cdots y_{rc}
\end{array}
$$

In our experiment y_{ij} would represent the time required by the ith operator to complete the problems on the jth machine. We shall let Y_{ij} represent a random variable for which y_{ij} is an observed value, and we shall assume that the random variables Y_{ij}, $i = 1, \ldots, r$, $j = 1, \ldots, c$, are independently normally distributed with a common variance σ^2 and with means given by the linear model

(16) $E(Y_{ij}) = \mu + a_i + b_j, \qquad i = 1, \ldots, r, \qquad j = 1, \ldots, c.$

This model implies that repetitions of our experiment would give rise to values of y_{ij} in the i, j cell of Table 2 that can be treated as the observed values of a normal random variable, and that the variability of such values about their mean would be the same for all cells. Furthermore, the mean for any such cell will depend linearly upon a factor that is a function of the operator's skill and on a factor that is a function of the brand of machine. Since μ includes all other factors that contribute to the total mean time required to solve the problems, the parameters a_i and b_j need to measure only the differences in operator skill and machine brand, respectively; therefore we shall assume that

(17) $\displaystyle\sum_{i=1}^{r} a_i = 0 \quad \text{and} \quad \sum_{j=1}^{c} b_j = 0.$

Thus a_i will be positive if the ith operator has a larger total time over all machines than the mean for the group and will be a negative number if the converse is true. Similarly, b_j will be positive if the jth machine requires more total operator time than the mean time for all machines.

The linearity assumption made in (16) can be quite unrealistic in some experiments. For example, suppose two different chemical compounds are mixed in varying amounts and applied to plots of growing vegetables as possible fertilizer mixtures. For some chemical compounds it might happen that each chemical by itself would be beneficial, but when mixed they would not be beneficial. Our model implies that each variable effect operates independently of the other variable effect, or equivalently, that there is no interaction between the variables in their effect upon Y.

For the purpose of demonstrating that this model is a special case of our general linear regression model, it is convenient to choose $n = rc$ and think of the variables Y_{ij} as being strung out in a line by taking a row at a time as follows: $Y_{11} \cdots Y_{1c} Y_{21} \cdots Y_{2c} \cdots Y_{r1} \cdots Y_{rc}$. The parameter μ and the a's and b's play the role of the β's in (8). Thus, the row vector $(\beta_1, \ldots, \beta_k)$ becomes the row vector $(\mu, a_1, \ldots, a_r, b_1, \ldots, b_c)$, where $k = r + c + 1$. It remains therefore to construct a set of x values that will enable (16) to be written in the form of (8). It will suffice to show what values are needed to produce $E(Y_{ij})$, and which are the values that would occur in row $(i - 1)c + j$ of the matrix X. These are easily seen to be the values $1, 0, \ldots, 0, 1, 0, \ldots, 0, 1, 0, \ldots, 0$ where the second 1 occurs in the $i + 1$ position and the third 1 occurs in the $r + 1 + j$ position. With this choice, (8) becomes

$$E(Y_{ij}) = \mu \cdot 1 + a_1 \cdot 0 + \cdots + a_i \cdot 1 + \cdots + a_r \cdot 0$$
$$+ b_1 \cdot 0 + \cdots + b_j \cdot 1 + \cdots + b_c \cdot 0,$$

which is the same as (16).

Since this analysis of variance model is a special case of the general linear regression model, our estimation problem is the same as for regression; therefore we may employ maximum likelihood methods if our normality assumption is justified, or least squares if we wish to make no distribution assumption. Since both methods yield the same equations, it suffices to solve the least squares problem. Unfortunately, the solution of the normal equations of least squares given by (11) cannot be applied here directly because our parameters are not all independent parameters. Since $\sum a_i = 0$ and $\sum b_j = 0$, there are only $r + c - 1$ independent parameters; therefore, in performing the calculus minimization of least squares, it is necessary to adjust for those dependencies. Our least squares problem is therefore the problem of choosing the parameters that minimize

$$G = \sum_{i=1}^{r} \sum_{j=1}^{c} (y_{ij} - \mu - a_i - b_j)^2,$$

subject to the restrictions $\sum a_i = 0$ and $\sum b_j = 0$. The problem can be reduced to one involving independent parameters by making the substitutions, say, $a_r = -\sum_{i=1}^{r-1} a_i$ and $b_c = -\sum_{j=1}^{c-1} b_j$ in G. The problem can also be solved by introducing two Lagrange multipliers λ_1 and λ_2 and solving the unrestricted problem of minimizing the function

$$H = G + \lambda_1 \sum_{i=1}^{r} a_i + \lambda_2 \sum_{j=1}^{c} b_j.$$

We shall use the latter method. First we calculate

$$\frac{\partial H}{\partial \mu} = \frac{\partial G}{\partial \mu} = -2 \sum_{i=1}^{r} \sum_{j=1}^{c} (y_{ij} - \mu - a_i - b_j) = 0.$$

Since $\sum_i a_i = 0$ and $\sum_j b_j = 0$, this equation yields the solution

$$\hat{\mu} = \sum_{i=1}^{r} \sum_{j=1}^{c} \frac{y_{ij}}{rc} = \bar{y}.$$

Inserting this value for μ in H and differentiating H with respect to a_s, where s is any index, will yield

$$\frac{\partial H}{\partial a_s} = -2 \sum_{j=1}^{c} (y_{sj} - \bar{y} - a_s - b_j) + \lambda_1 = 0.$$

Since $\sum_j b_j = 0$, this reduces to

(18) $$\sum_{j=1}^{c} (y_{sj} - \bar{y}) - ca_s = (1/2)\lambda_1.$$

This must hold for $s = 1, \ldots, r$. Summing both sides with respect to s will reduce the left side to zero because $\sum_s \sum_j y_{sj} = rc\bar{y}$ and $\sum_s a_s = 0$. As a result, $\lambda_1 = 0$, and therefore because s is an arbitrary index it follows from (18) by replacing s by i that

(19) $$\hat{a}_i = \bar{y}_{i \cdot} - \bar{y},$$

where $\bar{y}_{i \cdot} = \sum_j \frac{y_{ij}}{c}$. The dot subscript is a convenient device for indicating averaging with respect to that index. By symmetry, the least squares estimate of b_j will be given by

(20) $$\hat{b}_j = \bar{y}_{\cdot j} - \bar{y}.$$

In our contemplated experiment with calculating machines, $\hat{b}_j = \bar{y}_{\cdot j} - \bar{y}$ would serve as a measure of how much better or worse the jth machine is than the average of all machines and in which operator differences have been averaged out. Similarly, $\hat{a}_i$ would serve as a measure of the efficiency of the ith operator relative to the other operators of the experiment.

As an illustration, consider the problem of estimating the differential effects of car model and speed of driving on car mileage, if an experiment with five different model cars and three different speeds yielded the mileages shown in Table 3.

Table 3

	A	B	C	D	E	$\bar{y}_{i\cdot}$	$\bar{y}_{i\cdot} - \bar{y}$
I	19.8	18.5	23.0	18.0	24.5	20.8	1.5
II	18.7	17.8	22.4	16.8	23.1	19.8	.5
III	16.4	15.7	20.1	15.0	19.4	17.3	-2.0
$\bar{y}_{\cdot j}$	18.3	17.3	21.8	16.6	22.3	19.3	
$\bar{y}_{\cdot j} - \bar{y}$	-1.0	-2.0	2.5	-2.7	3.0		

Calculations here to the nearest decimal yielded the marginal means shown in Table 3 and a grand mean of $\bar{y} = 19.3$. Subtracting this mean from the marginal means produced the differential effects shown in the last row and last column. For these data there appears to be about as much variability in mileage due to car model as there is due to speed of driving.

The accuracy of the preceding estimates can be determined by means of confidence intervals, provided the value of σ is known. For example, to obtain a confidence interval for a_i it will suffice to calculate the mean and variance of $\bar{Y}_{i\cdot} - \bar{Y}$, because $\bar{Y}_{i\cdot} - \bar{Y}$ is a normal variable under the assumption that the Y_{ij} are normal variables. This is accomplished as follows.

$$\bar{Y}_{i\cdot} - \bar{Y} = \frac{1}{c} \sum_{j=1}^{c} Y_{ij} - \frac{1}{rc} \sum_{i=1}^{r} \sum_{j=1}^{c} Y_{ij}.$$

Since this is a linear combination of the Y_{ij}, which are independent normal variables, it is a normal variable. We know from Theorem 2 that the least squares estimators of the regression coefficients are unbiased; therefore since $\bar{Y}_{i\cdot} - \bar{Y} = \hat{a}_i$ is the least squares estimator of a_i,

$$E(\bar{Y}_{i\cdot} - \bar{Y}) = a_i.$$

Now write

$$\bar{Y}_{i\cdot} - \bar{Y} = \frac{1}{c} \sum_{k=1}^{c} Y_{ik} - \frac{1}{rc} \sum_{\alpha=1}^{r} \sum_{\beta=1}^{c} Y_{\alpha\beta}$$

$$= \sum_{k=1}^{c} Y_{ik} \left(\frac{1}{c} - \frac{1}{rc} \right) + \sum_{\alpha \neq i} \sum_{\beta=1}^{c} Y_{\alpha\beta} \left(-\frac{1}{rc} \right).$$

Then from the formula in Volume I for the variance of a linear combination of independent variables, it follows that

$$V(\bar{Y}_{i\cdot} - \bar{Y}) = \sum_{k=1}^{c} \sigma^2 \left(\frac{1}{c} - \frac{1}{rc} \right)^2 + \sum_{\alpha \neq i} \sum_{\beta=1}^{c} \sigma^2 \left(-\frac{1}{rc} \right)^2$$

$$= \sigma^2 \left[c \left(\frac{1}{c} - \frac{1}{rc} \right)^2 + (r - 1)c \left(-\frac{1}{rc} \right)^2 \right] = \sigma^2 \frac{r - 1}{rc}.$$

If σ is known, a 95 percent confidence interval for a_i is therefore given by

$$\hat{a}_i - 1.96\sigma_{\hat{a}_i} < a_i < \hat{a}_i + 1.96\sigma_{\hat{a}_i}.$$

When expressed in terms of the Y's, this becomes

$$\bar{Y}_{i.} - \bar{Y} - 1.96\sigma \sqrt{\frac{r - 1}{rc}} < a_i < \bar{Y}_{i.} - \bar{Y} + 1.96\sigma \sqrt{\frac{r - 1}{rc}}.$$

As a numerical illustration, we shall find a 95 percent confidence interval for a_3 for the problem related to Table 3. Here $r = 3$, $c = 5$, and $\hat{a}_3 = \bar{Y}_3. - \bar{Y} = -2.0$; hence the desired interval is

$$-2.0 - 1.96\sigma\sqrt{2/15} < a_3 < -2.0 + 1.96\sigma\sqrt{2/15}.$$

If the value of σ is not known, an approximation to this result can be obtained by replacing σ by its maximum likelihood estimate. Our likelihood function here is

$$L = \frac{\exp \left[-\frac{1}{2\sigma^2} \sum\limits_{i=1}^{r} \sum\limits_{j=1}^{c} (y_{ij} - \mu_{ij})^2 \right]}{(2\pi\sigma^2)^{rc/2}},$$

where $\mu_{ij} = \mu + a_i + b_j$. The usual calculus techniques will then give

$$(21) \qquad \hat{\sigma}^2 = \sum\limits_{i=1}^{r} \sum\limits_{j=1}^{c} (y_{ij} - \hat{\mu} - \hat{a}_i - \hat{b}_j)^2 / rc$$

$$= \sum\limits_{i=1}^{r} \sum\limits_{j=1}^{c} (y_{ij} - \bar{y}_{i.} - \bar{y}_{.j} + \bar{y})^2 / rc.$$

Calculations based on this formula for Table 3 yielded the estimate $\hat{\sigma} = .4$; therefore an approximation to the preceding confidence interval is the interval

$$-2.3 < a_3 < -1.7.$$

We shall not pursue the analysis of variance model further here because it will be discussed in considerable detail in the next chapter, where problems of testing hypotheses about regression parameters are studied, and where methods for finding confidence intervals that do not require approximations are derived.

BAYESIAN METHODS

If we assume as before that the random variables $Y_1, \ldots, Y_n$ are independently normally distributed with a common variance and means given by Formula (8), we can consider the Bayes problem of estimating the parameters $\beta_1, \ldots, \beta_k$, provided that we are prepared to assume that the β's possess a known density. It seldom occurs, however, that there is sufficient information available on the parameters of a regression function to justify assuming that they possess a known density. If such information is available, the earlier Bayesian methods can be applied to yield Bayesian estimates of the β's. The problem becomes quite complicated unless a simple loss function is selected. In particular, if we choose the loss function $\mathscr{L}(\beta, d) = \sum_{i=1}^{k} (d_i - \beta_i)^2$, the usual Bayesian techniques are easily applied and will yield the estimates given by

$$d_i = E(\beta_i \mid y_1, \ldots, y_n).$$

Because of their limited applicability, these methods will not be discussed further.

Exercises

1 Given the following set of points, find the equation of the least squares line that best fits them.

x	1	2	3	4	5
y	2	3	3	4	4

2 The following data are for the heights (x) and weights (y) of twelve college students. Find the equation of the least squares linear regression function for predicting weight from height.

x	63	63	65	66	68	69	70	70	71	72	72	74
y	124	126	133	140	164	154	161	164	150	172	184	210

3 Thirty samples of soil were analyzed for carbon content by two different methods. One method is quite accurate but expensive, whereas the other method is cheap but not very reliable. After subtracting a standard value from each measurement, calculations with the resulting thirty pairs of x, y sample values gave $\sum x_i = 60$, $\sum y_i = 90$, $\sum x_i^2 = 300$, $\sum y_i^2 = 750$, $\sum x_i y_i = 420$.
(a) Find the equation of the least squares line of y on x.
(b) Assuming that the regression line should pass through the origin, find the equation of the least squares line of y on x.

4 Derive the least squares equations for fitting a curve of the type $y = ax + \dfrac{b}{x}$ to a set of n points in the x, y plane. Is this a linear model?

5 Derive the least squares equations for fitting a curve of the type $y = axe^{-bx^2}$ to a set of n points in the x, y plane. Is this a linear model? If you chose log y as the variable to be predicted, what predictor would you use?

6 The pressure of a gas and its volume are related by an equation of the form $pv^a = b$. Take the logarithm of this relationship, define new variables and parameters, and estimate the values of a and b by least squares applied to the logarithmic relationship.

p	0.5	1.0	1.5	2.0	2.5	3.0
v	1.62	1.00	.75	.62	.52	.46

7 The following data give the velocity (y) of a river in feet per second corresponding to various depths expressed in terms of the ratio (x) of the measured depth to the depth of the river. Use least squares to fit a parabola $y = a + bx + cx^2$ to these points, choosing $x - .4$ rather than x as the independent variable.

x	0	.1	.2	.3	.4	.5	.6	.7	.8
y	3.20	3.23	3.25	3.26	3.25	3.23	3.18	3.13	3.06

8 Given $f(x, y) = 2, 0 \leq x \leq 1, 0 \leq y \leq x$, find the equation of the regression curve of y on x.

9 Given $f(x, y) = xe^{-x(y+1)}$, $x \geq 0$, $y \geq 0$, find the equation of the regression curve of y on x.

10 If X and Y are independent random variables, show that the curve of regression, if it exists, will be a horizontal straight line. What is the equation of the line?

11 Let Y, X_1, and X_2 denote the amount of hay in units of 100 pounds per acre, the spring rainfall in inches, and the accumulated temperature above 45°F in the spring, respectively. Data accumulated over several years in England yielded the following sample values: $\bar{Y} = 28, \bar{X}_1 = 4.9, \bar{X}_2 = 590, \sum x_1 y/n = 3.87, \sum x_2 y/n = -150, \sum x_1 x_2/n = -52, \sum x_1^2/n = 1.2, \sum x_2^2/n = 7,220$. Use these results to find the equation of the least squares linear regression function of Y on X_1 and X_2. Here a lower case letter denotes a variable measured from its empirical mean. Thus, $x_1 = X_1 - \bar{X}_1$, etc.

12 The following sample values, based on a sample of size 10, are for the variables honor points (Y), general intelligence test score (X_1), and hours of study (X_2). Find the equation of the least squares linear regression function of y on x_1 and x_2, if each variable has been measured from its sample mean so that $\sum y = \sum x_1 = \sum x_2 = 0$. Here $\sum x_1 y = 106, \sum x_2 y = 22, \sum x_1 x_2 = 33, \sum x_1^2 = 250, \sum x_2^2 = 36$.

13 Work Exercise 1 by using (11).

14 Work Exercise 2 by using (11).

15 Work Exercise 3 by using (11).

16 Show that the set of equations (7) is the same as the set of equations given by (10).

17 Employ (15) to obtain the variances of the estimates of the regression coefficients for Exercise 1 assuming that σ^2 is given. Use the results of Exercise 13 here.

18 Do as in Exercise 17 with respect to Exercises 2 and 14.

19 Do as in Exercise 17 with respect to Exercises 3 and 15.

20 The following data are for the yield of a vegetable crop on an experimental plot of ground (Y), the units of fertilizer added to the plot (X_1), and the number of irrigations (X_2).
   (a) Find the equation of the least squares linear regression function by means of (11).
   (b) Find the variances of the estimates of the regression coefficients as functions of σ^2.
   (c) Comment on the effectiveness of fertilizer and irrigation on yield.

y	24	23	27	26	25	26	29	27	31
x_1	0	0	0	1	1	1	2	2	2
x_2	2	4	6	2	4	6	2	4	6

21 The following data are for grade point averages (Y), intelligence test scores (X_1), and reading rates (X_2) of a set of students. The original test scores have been divided by 10 and rounded off. Use these data to find the equation of the least squares linear regression function of y on x_1 and x_2.

y	.6	.2	.0	1.0	.4	1.0	.0	2.4	2.6	1.8
x_1	15	21	17	14	13	22	12	29	23	18
x_2	18	45	24	30	26	25	20	42	44	24

y	1.4	.2	.4	1.4	.8	.8	2.2	1.2	1.4	2.6
x_1	23	13	18	23	18	21	23	15	20	27
x_2	44	30	28	50	32	20	28	20	30	40

Calculations give: $\sum x_1 = 385$, $\sum x_2 = 620$, $\sum y = 22.4$, $\sum x_1^2 = 7{,}841$, $\sum x_2^2 = 21{,}030$, $\sum x_1 x_2 = 12{,}509$, $\sum x_1 y = 488.4$, $\sum x_2 y = 765.6$, and

$$(X'X)^{-1} = \begin{bmatrix} .93908 & -.03802 & -.00507 \\ & .00404 & -.00128 \\ & & .00096 \end{bmatrix}$$

22 Use (15) to obtain the variances of the estimates of the regression coefficients for Exercise 12 assuming that σ^2 is given.

23 Use (15) to obtain the variances of the estimates of the regression coefficients for Exercise 21 assuming that σ^2 is given. Use the results of the calculations for Exercise 21.

24 Use $s^2 = \sum (y_i - y_i')^2/(n - 3)$ to estimate σ^2, where y_i' is the regression function value corresponding to the observed value y_i, for Exercise 20. This can be shown to be an unbiased estimate of σ^2. Replace σ^2 by s^2 in Exercise 20 to obtain estimates of the variances of the empirical regression coefficients. On the basis of these values and the usual normality assumption, find approximate 95 percent confidence limits for the coefficients.

25 For the regression line $y = \hat{a} + \hat{b}(x - \bar{x})$ with $\hat{a}$ and $\hat{b}$ given by (4), show by means of the covariance matrix that
(a) $V(\hat{a}) = \sigma^2/n$ and $V(\hat{b}) = \sigma^2/\sum (x_i - \bar{x})^2$.
(b) $\hat{a}$ and $\hat{b}$ are uncorrelated.

26 Use the calculations of Exercise 25 to find a formula for the variance of the random variable $Z = \hat{a} + \hat{b}(x - \bar{x})$. Graph this variance as a function of x to show how the precision for estimating a simple linear regression function value corresponding to a selected x value decreases rapidly as the x value moves away from $\bar{x}$.

27 On the basis of the formula obtained in Exercise 26, what values of $x_1, \ldots, x_n$ would you choose if you wished to estimate $E(Y \mid x)$ with maximum precision at an arbitrary x value and you are restricted to the interval $[-1, 1]$ for your x_i values?

28 Two polynomials $P_i(x)$ and $P_j(x)$ of degrees i and j, respectively, are said to be orthogonal on a set of points $x_1, \ldots, x_n$ provided that $\sum_{k=1}^{n} P_i(x_k)P_j(x_k) = 0, i \neq j$. The polynomial is said to be normalized on this set of points if $\sum_{k=1}^{n} P_i^2(x_k) = 1$. For the set of points $x = 0, 1, 2, 3$, find an orthogonal normalized (orthonormal) set of polynomials $P_0(x)$, $P_1(x)$, and $P_2(x)$.

29 Assuming the properties in Exercise 28, find the least squares estimates of the coefficients in the polynomial regression function given by $y = a_0 + a_1 P_1(x) + \cdots + a_r P_r(x)$, and show that the estimate of a_i is the same regardless of the degree of the polynomial (r) provided $i < r$. This implies that it is possible to add higher degree terms after a least squares fitting has already been made without changing the estimates already obtained, provided orthonormal polynomials with respect to the data points are used.

30 Carry out the minimization of G in Section 4.7 without the use of Lagrange multipliers by substituting $a_r = -\sum_1^{r-1} a_i$ and $b_c = -\sum_1^{c-1} b_j$ in G and calculating the necessary partial derivatives.

31 The following data give the gains of four types of turkeys fed three different rations over a period of several months. Find estimates of the differential effects of (a) rations, (b) types. Comment on these values. Save your answers for use in the next chapter.

Type

		I	II	III	IV
	A	6	15	10	12
Ration	B	14	16	14	18
	C	8	14	10	15

32 The following data give the number of units of work done per day by five workmen using four different types of machines. Each workman operated each type of machine for one day. Find estimates of the differential effects of (a) machine type, (b) workman skill. Comment on these values. Save your answers for use in the next chapter.

Type

		I	II	III	IV
	1	40	40	48	36
	2	40	42	50	48
Workman	3	35	37	45	32
	4	42	36	48	30
	5	36	40	50	40

33 Calculate an estimate of σ^2 for Exercise 31 by using the formula $s^2 = \sum\sum (y_{ij} - \bar{y}_{i\cdot} - \bar{y}_{\cdot j} + \bar{y})^2/(r-1)(c-1)$ in place of (21). It can be shown that S^2 is an unbiased estimate of σ^2, whereas $\hat{\sigma}^2$ is not. Use the identity

$$\sum\sum (y_{ij} - \bar{y})^2 = \sum\sum (\bar{y}_{i\cdot} - \bar{y})^2 + \sum\sum (\bar{y}_{\cdot j} - \bar{y})^2$$
$$+ \sum\sum (y_{ij} - \bar{y}_{i\cdot} - \bar{y}_{\cdot j} + \bar{y})^2$$

to compute the last term from the other terms, which are easy to calculate when the computations of Exercise 31 are available.

34 Calculate an estimate of σ^2 for Exercise 32 by using the formula and techniques of Exercise 33.

35 Demonstrate the validity of the identity employed in Exercise 33 by writing

$$\sum\sum (y_{ij} - \bar{y})^2 = \sum\sum [(\bar{y}_{i\cdot} - \bar{y}) + (\bar{y}_{\cdot j} - \bar{y})$$
$$+ (y_{ij} - \bar{y}_{i\cdot} - \bar{y}_{\cdot j} + \bar{y})]^2,$$

expanding it as a trinomial, and showing that all cross product terms vanish when summed in the proper order.

36 Show that the variable $\bar{Y}_{.j} - \bar{Y}$ is a normal variable. Obtain a formula for the variance of this variable in terms of σ^2 by expressing $\bar{Y}_{.j} - \bar{Y}$ in terms of the Y_{ij} and using the fact that the Y_{ij} are independent variables.

37 Use the results of Exercises 33 and 36 to find an approximate 90 percent confidence interval for b_1 of Exercise 31 by substituting s^2 for σ^2.

38 Show that the variables $\bar{Y}_{.j}$, $j = 1, \ldots, c$, are independent normal variables with the same mean and variance if $b_j = 0$, $j = 1, \ldots, c$.

39 Use the results of Exercise 38 to show that $r \sum_{j=1}^{c} (\bar{Y}_{.j} - \bar{Y})^2/\sigma^2$ possesses a chi-square distribution with $c - 1$ degrees of freedom if $b_j = 0$, $j = 1, \ldots, c$.

40 Show that the variable $Y_{ij} - \bar{Y}_{i.} - \bar{Y}_{.j} + \bar{Y}$ is a normal variable with zero mean.

41 Show that the variance of the variable in Exercise 40 is independent of i and j. Find a formula for it.

42 Use the results of Exercise 41 to show that

$$E \sum \sum (Y_{ij} - \bar{Y}_{i.} - \bar{Y}_{.j} + \bar{Y})^2 = \sigma^2(r - 1)(c - 1),$$

and hence that S^2 of Exercise 33 is an unbiased estimate of σ^2.

43 Write down the matrix X that expresses (16) in the form $EY = X\beta$, where β' is the vector $(\mu, a_1, \ldots, a_r, b_1, \ldots, b_c)$.

44 Show that $\sum (y_i - y_i')^2 = \sum \left[y_i - \bar{y} - r \frac{s_y}{s_x} (x_i - \bar{x}) \right]^2 = ns_y^2(1 - r^2)$.

45 Using the independence of $\hat{a}$ and $\hat{b}$ as shown in Exercise 25, show that $\hat{a}$ and $\hat{b}$ possess minimum variance among all linear unbiased estimators of α and β, and thus justify the least squares property stated at the end of Section 4.5 for simple linear regression.

5

Linear Models— Testing

In the preceding chapter the parameters of a general linear regression model were estimated by the method of maximum likelihood under the assumption that the random variables $Y_1, \ldots, Y_n$ possessed a normal distribution, and by the method of least squares when there was no such distribution assumption. Although the two methods produced the same estimates, it was pointed out that a distribution assumption is necessary if probability statements are to be made concerning the accuracy of those estimates, or if tests of hypotheses about the parameters are desired. Since we will be testing hypotheses in this chapter, we shall impose the same normality assumption as we did in the preceding chapter when estimating parameters by the method of maximum likelihood. The mathematical development here will be somewhat sophisticated and requires extensive use of the geometry of vector spaces. The result of the development will, however, be a test of such generality that it is capable of solving a large share of the important testing problems that arise in statistical practice. In view of the importance of this test, a heavy dose of theory is more than justified. In this connection, the matrix notation that was introduced in the preceding chapter is well nigh indispensible here and will be used almost exclusively. It will be found in Section 4.4. A review of the matrix methods needed in this development can be found in the appendix.

5.1. The general linear hypothesis

We shall assume that the random variables $Y_1, \ldots, Y_n$ are independently normally distributed with a common variance σ^2 and with means given by the formula $E(Y) = X\beta$. This formula is the matrix version of the general linear regression model given by Formulas (8) and (9) of Section 4.4. The density of the Y's is therefore given by

$$(1) \quad f(y_1, \ldots, y_n) = \frac{\exp\left[-\frac{1}{2\sigma^2} \sum_{i=1}^{n} (y_i - \beta_1 x_{i1} - \cdots - \beta_k x_{ik})^2 \right]}{(2\pi)^{n/2} \sigma^n}$$

140

$$= \frac{\exp\left[-\frac{1}{2\sigma^2}(y - X\beta)'(y - X\beta)\right]}{(2\pi)^{n/2}\sigma^n}.$$

All the interesting problems of testing hypotheses about the β's can be treated under one general setup; therefore we shall solve this general problem and then apply it to important practical special cases. The general linear hypothesis that we wish to test is expressible in the form

(2) $$H_0: A\beta = 0,$$

where A is a $q \times k$ matrix of full rank q, where $q \leq k$. This hypothesis therefore states that the β's satisfy q independent homogeneous linear restrictions. When written out in full, H_0 will assume the form

(3)
$$a_{11}\beta_1 + \cdots + a_{1k}\beta_k = 0$$
$$\vdots \qquad \qquad \vdots$$
$$a_{q1}\beta_1 + \cdots + a_{qk}\beta_k = 0.$$

As an illustration, suppose our regression function is chosen as $E(Y) = \beta_1 + \beta_2 x + \beta_3 x^2$, and we wish to test the hypothesis that a linear function will suffice, which means that $\beta_3 = 0$. Here A will be the 1×3 matrix given by $A = (0, 0, 1)$. We are not able to test directly a hypothesis of the form $H_0: \beta_3 = c$ where $c \neq 0$; however, it will be shown later that any such problem can be reduced to one for which $c = 0$. As a second illustration, suppose our regression function is of the form $E(Y) = \beta_1 x_1 + \beta_2 x_2 + \beta_3 x_3$, and we wish to test the hypothesis that $\beta_1 = \beta_2 = \beta_3$. Then A may be chosen to be the 2×3 matrix given by

$$A = \begin{bmatrix} 1 & -1 & 0 \\ 1 & 0 & -1 \end{bmatrix}.$$

The unknown parameters in our model are the β's and σ; therefore any hypothesis of the form (2) is a composite hypothesis. Since we are employing likelihood ratio tests for testing composite hypotheses, the problem here is to derive the likelihood ratio test for testing this general linear hypothesis.

The derivation of the test is quite lengthy and may involve some mathematics that is not familiar to some students; therefore the result of the development will be expressed in the form of a theorem and an indication of its applicability given before the proof is presented. The student without the necessary mathematical background may skip the proof and content himself with understanding and applying this basic theorem.

In the following theorem the random variable F is understood to be the random variable denoted by $F(k_1, k_2)$ in Section 6.6 of Volume I and which was defined there as the ratio of two independent chi-square variables after they had been divided by their degrees of freedom.

Theorem 1 *Assume that the random variables $Y_1, \ldots, Y_n$ are independently normally distributed with a common variance σ^2 and with means given by $\mu = E(Y) = X\beta$, where X is an $n \times k$ matrix of rank $k < n$. Then the likelihood ratio test for testing the hypothesis $H_0: A\beta = 0$, where A is a $q \times k$ matrix of rank $q \leq k$, is equivalent to a test based on the random variable F that uses as critical region $F \geq F_0$, where $P(F \geq F_0) = \alpha$ and where F is given by*

$$(4) \qquad F = \frac{\sum_{i=1}^{n} (y_i - \hat{\hat{\mu}}_i)^2 - \sum_{i=1}^{n} (y_i - \hat{\mu}_i)^2}{\sum_{i=1}^{n} (y_i - \hat{\mu}_i)^2} \frac{n - k}{q}.$$

The values $\hat{\mu}_i$ and $\hat{\hat{\mu}}_i$ are those that minimize $\sum_{i=1}^{n} (y_i - \mu_i)^2$ when μ satisfies the respective restrictions (a) $\mu = X\beta$, and (b) $\mu = X\beta$ and $A\beta = 0$. The degrees of freedom of the F variable are q and $n - k$.

To indicate how this theorem can be applied in treating a standard testing problem, consider the earlier illustration of testing whether $\beta_3 = 0$ in the model $E(Y) = \beta_1 + \beta_2 x + \beta_3 x^2$. Here $\hat{\mu}_i = \hat{\beta}_1 + \hat{\beta}_2 x_i + \hat{\beta}_3 x_i^2$, where $\hat{\beta}_1, \hat{\beta}_2$, and $\hat{\beta}_3$ are the least squares estimates of those parameters, namely those that minimize $\sum_{i=1}^{n} (y_i - \beta_1 - \beta_2 x_i - \beta_3 x_i^2)^2$. Similarly, $\hat{\hat{\mu}}_i = \hat{\hat{\beta}}_1 + \hat{\hat{\beta}}_2 x_i$, where $\hat{\hat{\beta}}_1$ and $\hat{\hat{\beta}}_2$ are the estimates that minimize $\sum_{i=1}^{n} (y_i - \beta_1 - \beta_2 x_i)^2$. Since $k = 3$ and $q = 1$ for this problem, the F value when written out in full will be

$$F = \frac{\sum_{i=1}^{n} (y_i - \hat{\hat{\beta}}_1 - \hat{\hat{\beta}}_2 x_i)^2 - \sum_{i=1}^{n} (y_i - \hat{\beta}_1 - \hat{\beta}_2 x_i - \hat{\beta}_3 x_i^2)^2}{\sum_{i=1}^{n} (y_i - \hat{\beta}_1 - \hat{\beta}_2 x_i - \hat{\beta}_3 x_i^2)^2} \frac{n - 3}{1}.$$

Problems of the preceding type will be discussed in considerable detail in the section on applications.

The proof of the preceding theorem consists of two parts. The first part derives the likelihood ratio test and shows that it is equivalent to a test based on the F statistic given in the statement of the theorem. The second part shows that this F statistic does possess the familiar F distribution as claimed. The first part of the proof is relatively easy and should be studied by all students. The second part requires some knowledge of vector spaces, and therefore it should be passed over by students who do not possess this necessary algebraic background. These students should move to section 5.2 where another useful theorem is stated and proved. The proof of that theorem also requires an algebraic background and also may be passed over.

5.1.1. Likelihood ratio test. Since we are dealing with a sample of size one with respect to the random variables $Y_1, \ldots, Y_n$, the likelihood function $L(\theta)$ is the density given by (1), but for convenience it will be expressed in the following more compact form.

$$(5) \quad L(\theta) = \frac{\exp\left[-\dfrac{1}{2\sigma^2}\sum_{i=1}^{n}(y_i - \mu_i)^2\right]}{(2\pi)^{n/2}\sigma^n} = \frac{\exp\left[-\dfrac{1}{2\sigma^2}(y - \mu)'(y - \mu)\right]}{(2\pi)^{n/2}\sigma^n}.$$

Let $\hat{\mu} = X\hat{\beta}$, where $\hat{\beta}$ denotes the maximum likelihood estimate of the vector parameter β, and let $\hat{\sigma}$ denote the maximum likelihood estimate of σ. From earlier work we know that

$$(6) \qquad\qquad \hat{\sigma}^2 = \frac{1}{n}\sum_{i=1}^{n}(y_i - \hat{\mu}_i)^2.$$

Hence,

$$L(\hat{\theta}) = \frac{e^{-(n/2)}}{(2\pi)^{n/2}(\hat{\sigma})^n}.$$

Let $\hat{\hat{\mu}} = X\hat{\hat{\beta}}$ and let $\hat{\hat{\sigma}}$ denote the corresponding estimates when H_0 is assumed to be true. Then

$$L(\hat{\theta}_0) = \frac{e^{-(n/2)}}{(2\pi)^{n/2}(\hat{\hat{\sigma}})^n}.$$

As a result, the likelihood ratio $\lambda = L(\hat{\theta}_0)/L(\hat{\theta})$ becomes

$$\lambda = (\hat{\sigma}/\hat{\hat{\sigma}})^n.$$

The critical region $\lambda \leq \lambda_0$ is equivalent to the region $\lambda^{-(2/n)} \geq \lambda_0^{-(2/n)}$, which is of the form

$$\hat{\hat{\sigma}}^2/\hat{\sigma}^2 \geq c.$$

In view of (6), this critical region can be expressed as

$$\frac{\sum_{i=1}^{n}(y_i - \hat{\hat{\mu}}_i)^2}{\sum_{i=1}^{n}(y_i - \hat{\mu}_i)^2} \geq c.$$

For theoretical reasons it is more convenient to work with the equivalent critical region given by

$$(7) \qquad \frac{\sum_{i=1}^{n}(y_i - \hat{\hat{\mu}}_i)^2 - \sum_{i=1}^{n}(y_i - \hat{\mu}_i)^2}{\sum_{i=1}^{n}(y_i - \hat{\mu}_i)^2} \geq c - 1.$$

Under our normality assumption the maximum likelihood estimates of the μ's are identical with the least squares estimates; therefore the left side of (7) is the same as the corresponding expression in Theorem 1.

5.1.2.   Distribution derivation.   The problem now is to show that this critical region is equivalent to one based on an F variable. We shall do this by finding the distribution of the random variable corresponding to the left side of (7) by a combination of algebraic and geometrical methods.

For this purpose let the sample point $y_1, \ldots, y_n$ be represented as a vector in n dimensional sample space as shown in Figure 1. For convenience of reference, the matrix X is displayed again, with its column vectors denoted by $\xi_1, \ldots, \xi_k$.

$$X = \begin{array}{c} \begin{array}{cccc} \xi_1 & \xi_2 & \cdots & \xi_k \end{array} \\ \begin{bmatrix} x_{11} & x_{12} & \cdots & x_{1k} \\ \vdots & \vdots & & \vdots \\ x_{n1} & x_{n2} & \cdots & x_{nk} \end{bmatrix} \end{array}.$$

Since the mean vector μ is given by $\mu = X\beta$, it can be expressed in terms of these column vectors as

(8) $$\mu = \beta_1 \xi_1 + \cdots + \beta_k \xi_k.$$

The vectors $\xi_1, \ldots, \xi_k$ will generate a k dimensional subspace of our sample space because X is assumed to be of full rank $k < n$, and the ξ's therefore are linearly independent. This subspace will be denoted by $\mathscr{L}$. Because of (8) the vector μ is restricted to lie somewhere in $\mathscr{L}$. When the hypothesis H_0 is assumed to be true, there are q independent homogeneous linear restrictions on the β's as shown in (3), which can be used to eliminate q of the β's in (8) and reduce it to the form

$$\mu = \beta_1' \eta_1 + \cdots + \beta_{k-q}' \eta_{k-q}.$$

Here $\beta_1', \ldots, \beta_{k-q}'$ denote the β's that remain, and $\eta_1, \ldots, \eta_{k-q}$ represent linear combinations of the ξ's that result from the elimination. The η's are clearly linearly independent vectors because only η_i involves ξ_i, and therefore they generate a $k - q$ dimensional subspace of $\mathscr{L}$, which will be denoted by $\mathscr{L}_0$. Thus, when H_0 is assumed to be true, the vector μ is restricted to lie somewhere in $\mathscr{L}_0$.

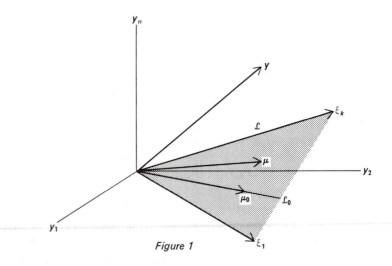

Figure 1

The subspaces $\mathscr{L}$ and $\mathscr{L}_0$ together with corresponding typical mean vectors, denoted by μ and μ_0, are shown in Figure 1. In that sketch $\mathscr{L}$ is represented by the two dimensional plane determined by the column vectors ξ_1 and ξ_k, and only part of which is shown, and $\mathscr{L}_0$ is represented by a line lying in that plane. Although the sketch is only for $n = 3, k = 2$, and $q = 1$ because of our limitations in sketching higher dimensional spaces, it should help explain the geometry of the problem for the general case.

Now let $\alpha_1, \ldots, \alpha_n$ denote a set of normal orthogonal vectors that will span n dimensional sample space, with the vectors $\alpha_1, \ldots, \alpha_k$ spanning $\mathscr{L}$, and with the first $k - q$ of them, namely $\alpha_1, \ldots, \alpha_{k-q}$, spanning $\mathscr{L}_0$. Recall from algebra that orthonormal vectors satisfy the restrictions

$$(9) \qquad \alpha_i'\alpha_j = 0, \qquad i \neq j, \qquad \text{and} \qquad \alpha_i'\alpha_i = 1,$$

where the prime here denotes the transpose of the column vector α_i. Let $z_1, \ldots, z_n$ denote the coordinates of y with respect to these new basis vectors. This implies that the vector y can be expressed in the form

$$(10) \qquad y = \sum_{j=1}^{n} z_j\alpha_j.$$

Similarly, let $v_1, \ldots, v_n$ denote the coordinates of μ with respect to our new basis. Since μ is restricted to lie in $\mathscr{L}$ and $\mathscr{L}$ is spanned by the vectors $\alpha_1, \ldots, \alpha_k$, the vector μ must be of the form

$$(11) \qquad \mu = \sum_{j=1}^{k} v_j\alpha_j.$$

The purpose of introducing this new basis for sample space is to simplify the sums of squares occurring in (7). This will be accomplished as follows. Because of (10) and (11), the vector $y - \mu$ can be written as

$$y - \mu = \sum_{j=1}^{k} (z_j - v_j)\alpha_j + \sum_{j=k+1}^{n} z_j\alpha_j.$$

As a result, it is easily shown by applying properties (9) that the basic sum of squares $\sum_{i=1}^{n} (y_i - \mu_i)^2 = (y - \mu)'(y - \mu)$ will assume the form

$$(12) \qquad (y - \mu)'(y - \mu) = \sum_{j=1}^{k} (z_j - v_j)^2 + \sum_{j=k+1}^{n} z_j^2.$$

The vector μ given by (11) will minimize this sum of squares if we choose $v_j = z_j, j = 1, \ldots, k$. This choice is permissible because as μ ranges over $\mathscr{L}$ the v's will range over all real numbers. Thus, from (11) we obtain the result that

$$(13) \qquad \hat{\mu} = \sum_{j=1}^{k} z_j\alpha_j,$$

and from (12) that

$$\sum_{i=1}^{n} (y_i - \hat{\mu}_i)^2 = \sum_{j=k+1}^{n} z_j^2.$$

When H_0 is assumed to be true, μ must lie in $\mathscr{L}_0$. But $\mathscr{L}_0$ is spanned by $\alpha_1, \ldots, \alpha_{k-q}$; therefore the expression for μ corresponding to (11) will involve only the first $k - q$ of those α's. By the same reasoning as before, the sum of squares corresponding to (12) will be minimized by choosing $v_j = z_j, j = 1, \ldots, k - q$, from which it follows that

(14) $$\hat{\hat{\mu}} = \sum_{j=1}^{k-q} z_j \alpha_j$$

and

$$\sum_{i=1}^{n} (y_i - \hat{\hat{\mu}}_i)^2 = \sum_{j=k-q+1}^{n} z_j^2.$$

In terms of the z's, the critical region (7) is therefore given by

$$\frac{\sum_{j=k-q+1}^{n} z_j^2 - \sum_{j=k+1}^{n} z_j^2}{\sum_{j=k+1}^{n} z_j^2} \geq c - 1.$$

This reduces to

(15) $$\frac{\sum_{j=k-q+1}^{k} z_j^2}{\sum_{j=k+1}^{n} z_j^2} \geq c - 1.$$

Now consider the relationship between the y's and the z's as given by (10). From (9) and (10) it follows that

(16) $$\alpha_i' y = \sum_{j=1}^{n} z_j \alpha_i' \alpha_j = z_i.$$

Hence the vector z is expressed in terms of the vector y by the relation

$$z = Py,$$

where P is the matrix with the vector α_i' as its ith row. Thus,

$$P = \begin{bmatrix} \alpha_1' \\ \vdots \\ \alpha_n' \end{bmatrix}.$$

From properties (9) it follows that $PP' = I$, and therefore that P is an orthogonal matrix.

We now introduce the random vector Z where $Z = PY$. From Volume I we know that the density of Z is given by the formula

(17) $$g(z_1, \ldots, z_n) = |P|^{-1} f(y_1, \ldots, y_n),$$

where the y's on the right must be expressed in terms of the z's by the relation $y = P^{-1}z = P'z$. Since P is an orthogonal matrix, it follows from $PP' = I$ and properties of determinants that $1 = |I| = |PP'| = |P| \, |P'| = |P|^2$, and hence that $|P| = 1$. Furthermore, since v is the mean vector μ in terms of the new basis, it follows that $v = P\mu$ or $\mu = P'v$; therefore

$$(y - \mu)'(y - \mu) = (P'z - P'v)'(P'z - P'v)$$

$$= (z - v)'PP'(z - v)$$

$$= (z - v)'(z - v).$$

These results together with (5) and (17) demonstrate that

$$g(z_1, \ldots, z_n) = \frac{\exp\left[-\frac{1}{2\sigma^2}(z - v)'(z - v)\right]}{(2\pi)^{n/2}\sigma^n} = \frac{\exp\left[-\frac{1}{2\sigma^2}\sum_{i=1}^{n}(z_i - v_i)^2\right]}{(2\pi)^{n/2}\sigma^n}.$$

This proves that the random variables $Z_1, \ldots, Z_n$ are independently normally distributed with the common variance σ^2 and with means given by $v = P\mu$.

Since v is the vector μ with respect to the new basis, and μ is restricted to lie in $\mathscr{L}_0$ when H_0 is assumed to be true, it follows that all except the first $k - q$ coordinates of v must be zero; therefore $v_{k-q+1} = \cdots = v_n = 0$ when H_0 is true. Thus,

$$(18) \quad g(z_1, \ldots, z_n) = \frac{\exp\left[-\frac{1}{2\sigma^2}\left[\sum_{i=1}^{k-q}(z_i - v_i)^2 + \sum_{i=k-q+1}^{n} z_i^2\right]\right]}{(2\pi)^{n/2}\sigma^n}$$

when the hypothesis H_0 is assumed to be true.

It will be observed from (18) that $Z_{k-q+1}, \ldots, Z_n$ are independent normal variables with zero means and a common variance σ^2; consequently

$$\frac{\sum_{i=k-q+1}^{k} Z_i^2}{\sigma^2} \quad \text{and} \quad \frac{\sum_{i=k+1}^{n} Z_i^2}{\sigma^2}$$

will possess independent chi-square distributions with q and $n - k$ degrees of freedom, respectively. Hence, it follows from Volume I, or an earlier exercise, that

$$\frac{\sum_{i=k-q+1}^{k} Z_i^2/q}{\sum_{i=k+1}^{n} Z_i^2/(n - k)}$$

will possess an F distribution with q and $n - k$ degrees of freedom when H_0 is true. In view of (15) the critical region of our likelihood ratio test is therefore equivalent to the region $F \geq (c - 1)(n - k)/q$. By choosing $(c - 1)(n - k)/q = F_0$, where $P(F \geq F_0) = \alpha$, the size of this critical region will be α. This demonstrates that the original likelihood ratio test is equivalent to a test based on the random variable F which chooses as its critical region those values of F satisfying $F \geq F_0$. In terms of the original variables the value of F is given by the left side of (7) after (7) has been multiplied by $(n - k)/q$. This completes the proof of our theorem. ∎

It is interesting to observe the geometry of the preceding derivation. The sum of squares $(y - \mu)'(y - \mu)$ represents the square of the length of the vector $y - \mu$. By analogy with three dimensions, it will be minimized if the vector $y - \mu$ is orthogonal to the subspace $\mathscr{L}$, which implies that $\hat{\mu}$ must be the projection of the vector y on $\mathscr{L}$. This is shown in Figure 2. Similarly, when μ is restricted to lie in $\mathscr{L}_0$, this squared length will be minimized when $y - \mu$ is orthogonal to $\mathscr{L}_0$, which implies that $\hat{\hat{\mu}}$ is the projection of the vector y on $\mathscr{L}_0$. This is also shown in Figure 2.

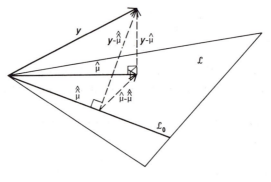

Figure 2

By means of (13) and (14), it is possible to express F in a very compact form, which will be found useful in some of the applications. Subtracting $\hat{\hat{\mu}}$ from $\hat{\mu}$ as they are expressed in (13) and (14) will give

$$\hat{\mu} - \hat{\hat{\mu}} = \sum_{j=k-q+1}^{k} z_j \alpha_j.$$

Because of the orthonormality of the α's, it then follows that

$$\sum_{i=1}^{n} (\hat{\mu}_i - \hat{\hat{\mu}}_i)^2 = (\hat{\mu} - \hat{\hat{\mu}})'(\hat{\mu} - \hat{\hat{\mu}}) = \sum_{j=k-q+1}^{k} z_j^2.$$

In view of the equivalence of (7) and (15), this shows that

$$\sum_{i=1}^{n} (y_i - \hat{\hat{\mu}}_i)^2 - \sum_{i=1}^{n} (y_i - \hat{\mu}_i)^2 = \sum_{i=1}^{n} (\hat{\mu}_i - \hat{\hat{\mu}})^2.$$

As a result, we may write F in the more compact form

(19)
$$F = \frac{\sum_{i=1}^{n} (\hat{\mu}_i - \hat{\hat{\mu}}_i)^2}{\sum_{i=1}^{n} (y_i - \hat{\mu}_i)^2} \frac{n - k}{q}.$$

5.2.   Confidence intervals for regression coefficients

A special case of the general linear hypothesis test which is of considerable importance is that of testing whether a particular coefficient in the regression function, say β_j, has a given value β_j^0. By writing $\beta_j = b_j + \beta_j^0$ in the regression function and working with the random variables $Z_i = Y_i - \beta_j^0 x_{ij}$, the problem can be reduced to one of testing the hypothesis $H_0: b_j = 0$, which is of the required form for our theorem. This reduction is treated as a problem in the exercises.

Not only can our theorem be used to test whether β_j has a given value, but it can also be used to construct a confidence interval for β_j. The objective of this section is to show how this can be done.

For the regression function $\mu = X\beta$ we know from Section 4.4 that the least squares estimator $\hat{\beta}$ is given by the formula

$$(20) \qquad \hat{\beta} = (X'X)^{-1}X'Y.$$

We also know from Theorems 2 and 3 of Chapter 4 that $E\hat{\beta}_j = \beta_j$ and $V(\hat{\beta}_j) = \sigma^2 \sigma_{jj}$, where σ_{jj} is the jth diagonal element of the matrix $(X'X)^{-1}$. Since $Y_1, \ldots, Y_n$ are independent normal variables, and since $\hat{\beta}_j$ as given by (20) is a linear combination of those variables, it follows from the foregoing properties of $\hat{\beta}_j$ that $\hat{\beta}_j$ is a normal variable with mean β_j and variance $\sigma^2 \sigma_{jj}$. Consequently,

$$\frac{\hat{\beta}_j - \beta_j}{\sigma\sqrt{\sigma_{jj}}}$$

is a standard normal variable that can be used to obtain confidence intervals for β_j, if the value of σ is known. In practice, however, we seldom know the value of σ, and therefore a more sophisticated method is needed for such problems. A method that overcomes this difficulty is given by means of the following theorem.

Theorem 2 *The random variable*

$$(21) \qquad T = \frac{(\hat{\beta}_j - \beta_j)/\sqrt{\sigma_{jj}}}{\sqrt{\sum_{i=1}^{n}(Y_i - \hat{\mu}_i)^2/(n-k)}}$$

possesses a t distribution with $n - k$ degrees of freedom.

Theorem 2 may also be used to test the hypothesis $H_0: \beta_j = \beta_j^0$ where β_j^0 is a particular value of β_j. If the alternative hypothesis were $H_1: \beta_j \neq \beta_j^0$, we would choose the two-sided critical region given by $|t| > t_0$ where $P(|T| > t_0) = \alpha$ and where β_j is replaced by β_j^0. It can be shown by means of the material in the proofs of Theorems 1 and 2 that this test is exactly the same as the likelihood ratio test of Theorem 1 for this problem.

We shall now give a proof of Theorem 2 based on linear algebra techniques; therefore those students who skipped the proof of Theorem 1 should also skip this proof and move on to the applications in the next section.

In proving this theorem we first relabel the x's and β's if necessary so that β_j becomes β_k. Now let $\mathscr{L}$ be the subspace of n dimensional sample space spanned by $\xi_1, \ldots, \xi_k$, and let $\mathscr{L}_0$ be the subspace spanned by $\xi_1, \ldots, \xi_{k-1}$. Let $\alpha_1, \ldots, \alpha_n$ be an orthonormal basis of our sample space such that $\alpha_1, \ldots, \alpha_{k-1}$ is an orthonormal basis of $\mathscr{L}_0$ and $\alpha_1, \ldots, \alpha_k$ is an orthonormal basis of $\mathscr{L}$.

As in the proof of Theorem 1, we use the relation (16) to introduce random variables $Z_1, \ldots, Z_n$ defined by $Z_i = \alpha_i' Y$. We saw in that proof that $Z_1, \ldots, Z_n$ are independent normal variables with the common variance σ^2, that $Z_{k+1}, \ldots, Z_n$ have zero means, and that

$$\sum_{i=1}^{n} (Y_i - \hat{\mu}_i)^2 = \sum_{j=k+1}^{n} Z_j^2.$$

As a result

(22) $$\frac{1}{\sigma^2} \sum_{i=1}^{n} (Y_i - \hat{\mu}_i)^2 = \frac{1}{\sigma^2} \sum_{j=k+1}^{n} Z_j^2$$

possesses a chi-square distribution with $n - k$ degrees of freedom.

Since $\hat{\mu} = X\hat{\beta}$, it follows just as in (8) that

$$\hat{\mu} = \hat{\beta}_1 \xi_1 + \cdots + \hat{\beta}_k \xi_k.$$

Now the vectors $\xi_1, \ldots, \xi_{k-1}$ and the vectors $\alpha_1, \ldots, \alpha_{k-1}$ span the same space $\mathscr{L}_0$; therefore each of the vectors ξ_j, $j = 1, \ldots, k - 1$, can be expressed as a linear combination of the α's, say in the form, $\xi_j = c_{j1}\alpha_1 + \cdots + c_{jk-1}\alpha_{k-1}$. Since α_k is orthogonal to $\alpha_1, \ldots, \alpha_{k-1}$ because $\alpha_1, \ldots, \alpha_{k-1}, \alpha_k$ are an orthonormal set, it follows from this expression for ξ_j that

$$\xi_j' \alpha_k = 0, \qquad j = 1, \ldots, k - 1.$$

As a result it follows from the preceding expression for $\hat{\mu}$ that

$$\hat{\mu}' \alpha_k = \hat{\beta}_k \xi_k' \alpha_k.$$

From (13) we have the relation

$$\hat{\mu} = Z_1 \alpha_1 + \cdots + Z_k \alpha_k,$$

and hence from the orthonormality of the α's that

$$\hat{\mu}' \alpha_k = Z_k.$$

Equating these two expressions for $\hat{\mu}' \alpha_k$ gives

$$\hat{\beta}_k \xi_k' \alpha_k = Z_k.$$

Since Z_k has the variance $\sigma^2 > 0$, it follows that $\xi_k' \alpha_k \neq 0$ and therefore that we may divide through by the number $\xi_k' \alpha_k$ to obtain

$$\hat{\beta}_k = \frac{Z_k}{\xi'_k \alpha_k}.$$

From the independence of the variables $Z_k, Z_{k+1}, \ldots, Z_n$, it follows that $\hat{\beta}_k$, which is a function of the variable Z_k only, is independent of $\sum_{j=k+1}^{n} Z_j^2 / \sigma^2$; consequently it follows from (22) that

$$\frac{\hat{\beta}_k - \beta_k}{\sigma \sqrt{\sigma_{kk}}} \quad \text{and} \quad \sum_{i=1}^{n} (Y_i - \hat{\mu}_i)^2 / \sigma^2$$

are independent random variables. As was shown earlier, the first of these is a standard normal variable and the second is a chi-square variable with $n - k$ degrees of freedom. From Volume I we know that the ratio of a standard normal variable to the square root of an independent chi-square variable, when multiplied by the square root of the degrees of freedom of the chi-square variable, is a t variable. As a result

$$T = \frac{(\hat{\beta}_k - \beta_k)/\sqrt{\sigma_{kk}}}{\sqrt{\sum_{i=1}^{n} (Y_i - \hat{\mu}_i)^2 / (n - k)}}$$

possesses a t distribution with $n - k$ degrees of freedom. This completes the proof of the theorem. ∎

This theorem also enables us to justify the use of the t distribution in earlier chapters for testing normal means and differences of normal means. All that is needed here is to choose $k = 1$ and to choose X to be the $n \times 1$ matrix that has all its elements equal to 1. The vector $\mu = X\beta$ then becomes the vector that has all its components equal to the number β_1. As a result $\sum_{i=1}^{n} (y_i - \mu_i)^2$ becomes $\sum_{i=1}^{n} (y_i - \beta_1)^2$. The least squares estimate is then given by $\hat{\beta}_1 = \bar{y}$; consequently $\sum_{i=1}^{n} (y_i - \hat{\mu}_i)^2 = \sum_{i=1}^{n} (y_i - \bar{y})^2$. Since $\bar{Y}$ now plays the role of $\hat{\beta}_j$ in Theorem 2 and $(X'X)^{-1} = 1/n$ here, the formula of Theorem 2 will reduce to

$$T = \frac{\sqrt{n}(\bar{Y} - \mu)\sqrt{n - 1}}{\sqrt{\sum_{i=1}^{n} (Y_i - \bar{Y})^2}},$$

where μ is used in place of β_1 to represent $E(Y)$. This formula will be seen to be equivalent to Formula (32) of Chapter 3.

The preceding derivation shows that for a normal variable Y the sample mean $\bar{Y}$ and the sample variance $\sum_{i=1}^{n} (Y_i - \bar{Y})^2 / n$ are independent random variables. Furthermore, since $\sum_{i=1}^{n} (Y_i - \hat{\mu}_i)^2 / \sigma^2$ possesses a chi-square distribution with $n - k$ degrees of freedom, it follows that we also have the result that $\sum_{i=1}^{n} (Y_i - \bar{Y})^2 / \sigma^2$ possesses a chi-square distribution with $n - 1$ degrees of freedom. Both of these results were used without proof on several occasions in the earlier chapters.

APPLICATIONS

5.3. Simple linear regression

The problem of predicting the value of a random variable Y by means of the known value of a related variable X was discussed from the point of view of estimating such a predictor. In the case of a simple linear model our estimate was obtained by the technique of least squares. It was observed that the computations became somewhat simpler if our linear model was written in the form

(23) $$E(Y \mid x) = a + b(x - \bar{x}),$$

where $\bar{x}$ is the mean of the values of the x's of the sample.

Now suppose we have obtained least squares estimates $\hat{a}$ and $\hat{b}$ for this model. We may then wish to know how accurate these estimates are, or we may wish to know whether our estimates are compatible with previously postulated values of a and b. In particular, it is often of paramount interest to know whether the true value of b is zero, since if b is zero, knowing the value of x does not help us predict the corresponding value of Y. Because of the importance of this problem, we shall ignore a and concentrate on testing whether the value of b is zero.

If we were interested in testing some other value of b, call it b_0, we could reduce the problem to testing a zero value by letting $d = b - b_0$ and rewriting the equation of the straight line to be fitted to the data in the form

$$y = a + (d + b_0)(x - \bar{x}).$$

Then letting $z = y - b_0(x - \bar{x})$, our equation would reduce to

$$z = a + d(x - \bar{x}).$$

In this form testing $d = 0$ is equivalent to testing $b = b_0$. The only change required therefore is to replace y_i by $z_i = y_i - b_0(x_i - \bar{x})$ in the calculations.

Model (23) is a special case of $E(Y) = X\beta$, where $x_{i1} = 1$ and $x_{2i} = x_i - \bar{x}$, $i = 1, \ldots, n$, and where $\beta_1 = a$ and $\beta_2 = b$. The hypothesis $H_0: b = 0$ will become a special case of $H_0: A\beta = 0$ if we choose A to be the 1×2 matrix $A = (0, 1)$. To test this hypothesis it is first necessary to find the values of the $\hat{\mu}_i$ that minimize $\sum_{i=1}^{n} (y_i - \hat{\mu}_i)^2$. But $\hat{\mu}_i = \hat{\beta}_1 + \hat{\beta}_2(x_i - \bar{x})$, where $\hat{\beta}_1$ and $\hat{\beta}_2$ are the least squares estimates of β_1 and β_2. From Formulas (4), Chapter 4, these estimates were found to be

$$\hat{\beta}_1 = \bar{y} \quad \text{and} \quad \hat{\beta}_2 = \frac{\sum (x_i - \bar{x}) y_i}{\sum (x_i - \bar{x})^2}.$$

When H_0 is true, $\beta_2 = 0$, and therefore

$$\sum_{i=1}^{n} (y_i - \mu_i)^2 = \sum_{i=1}^{n} (y_i - \beta_1)^2.$$

The least squares estimate of β_1 here is obviously $\hat{\hat{\beta}}_1 = \bar{y}$. Application of our basic theorem given by (4) then yields the following F value

$$(24) \quad F = \frac{\sum_{i=1}^{n} (y_i - \bar{y})^2 - \sum_{i=1}^{n} (y_i - \hat{\beta}_1 - \hat{\beta}_2(x_i - \bar{x}))^2}{\sum_{i=1}^{n} (y_i - \hat{\beta}_1 - \hat{\beta}_2(x_i - \bar{x}))^2} \frac{n-2}{1}.$$

The degrees of freedom factor resulted from realizing that $k = 2$ and A is 1×2 so that $q = 1$.

As a numerical illustration, consider the problem of testing whether $b = 0$ for the straight line fitted to the set of points shown in Figure 1, Chapter 4. This is strictly a computational exercise because it seems clear from Figure 1 that $b = 0$ is an unrealistic hypothesis to be testing. From those earlier calculations

$$y_i' = \hat{\beta}_1 + \hat{\beta}_2(x_i - \bar{x}) = 52.1 + .56(x_i - 100).$$

It is therefore necessary to calculate $\sum (y_i - \bar{y})^2$ and $\sum (y_i - y_i')^2$. Such calculations yield the values

$$\sum (y_i - \bar{y})^2 = 2,574 \quad \text{and} \quad \sum (y_i - y_i')^2 = 1,170.$$

The corresponding F value is therefore given by

$$F = \frac{2,574 - 1,170}{1,170} 11 = 13.2.$$

From Table VII in the appendix it will be found that the .05 critical point for F based on 1 and 11 degrees of freedom is 4.84; therefore the hypothesis $H_0: b = 0$ is certainly rejected here.

Although (24) assumes that the least squares line to be fitted is of the form (23), it is not necessary to measure the x's from their mean. Formula (24) is equally valid if $\bar{x}$ is deleted; however, then $\hat{\beta}_1$ and $\hat{\beta}_2$ will denote the least squares estimates of a and b for the line $y = a + bx$.

5.4. Multiple linear regression

The problem that was studied in the preceding section will now be generalized to that of testing whether certain of the coefficients in a general regression model have zero values. As before, it is always possible to reformulate the problem to test nonzero values if that is required. We shall assume the normal regression model (1). By rearranging terms and relabeling variables, our hypothesis can be expressed in the form

$$H_0: \beta_{l+1} = \cdots = \beta_k = 0,$$

where l is some integer satisfying $0 \leq l \leq k - 1$.

Since $\mu = X\beta$, the least squares estimate of μ is given by $\hat{\mu} = X\hat{\beta}$, where $\hat{\beta}$ is the least squares estimate of β given by Formula (11), Chapter 4.

Let X_0 denote the submatrix of X obtained by deleting the last $k - l$ columns. When H_0 is true, $X\beta$ is the same as $X_0\beta$, because those last column terms would be multiplied by zero β values. The least squares estimate of μ is therefore given by $\hat{\hat{\mu}} = X_0\hat{\hat{\beta}}$, where $\hat{\hat{\beta}}$ is given by Formula (11), Chapter 4, with X replaced by X_0.

The matrix A that is needed to show that our problem is a special case of (4) is the $(k - l) \times k$ matrix given by

$$A = \begin{bmatrix} 0 & 0 \cdots 0 & 1 & 0 \cdots 0 \\ 0 & 0 \cdots 0 & 0 & 1 \cdots 0 \\ \vdots & \vdots & \vdots & \vdots & \vdots & \vdots \\ 0 & 0 \cdots 0 & 0 & 0 \cdots 1 \end{bmatrix}$$

where the 1 in the first row occurs in column $l + 1$. Since this matrix is of rank $k - l$, the degrees of freedom needed for the F test of (4) are $k - l$ and $n - k$. Our test is therefore based upon the following F value:

$$(25) \quad F = \frac{(y - X_0\hat{\hat{\beta}})'(y - X_0\hat{\hat{\beta}}) - (y - X\hat{\beta})'(y - X\hat{\beta})}{(y - X\hat{\beta})'(y - X\hat{\beta})} \frac{n - k}{k - l}.$$

There are a number of algebraic simplifications that could be made in this problem to ease the burden of computation. However, showing details of calculation would detract from our basic objective of showing how various important problems can be treated under one general formulation.

The problem of testing whether a polynomial regression function of degree $k - 1$ is one of only degree $l - 1$ can be solved directly by means of (25). It is merely necessary to replace x_{ij} by x_i^{j-1} in X and apply (25). This test is also capable of treating other types of regression functions, such as trigonometric, by making the proper substitution for x_{ij} in X.

As a numerical illustration, assume that the variables Y_i are independently normally distributed with a common unknown variance σ^2 and with means lying on a polynomial curve in x of degree three or less. Consider the problem of testing the hypothesis that the degree is 1 on the basis of the following data.

x	-3	-2	-1	0	1	2	3
y	1	4	5	5	6	7	9

First it is necessary to replace x_{ij} by x_i^{j-1}, $i = 1, \ldots, 7$, $j = 1, \ldots, 4$, in X. The least squares estimates $X\hat{\beta}$ and $X_0\hat{\hat{\beta}}$ needed in (25) can be obtained by employing Formula (11), Chapter 4, namely

$$\hat{\beta} = (X'X)^{-1}X'y.$$

Here

$$X' = \begin{bmatrix} 1 & 1 & 1 & 1 & 1 & 1 & 1 \\ -3 & -2 & -1 & 0 & 1 & 2 & 3 \\ 9 & 4 & 1 & 0 & 1 & 4 & 9 \\ -27 & -8 & -1 & 0 & 1 & 8 & 27 \end{bmatrix}.$$

Hence,

$$X'X = \begin{bmatrix} 7 & 0 & 28 & 0 \\ 0 & 28 & 0 & 196 \\ 28 & 0 & 196 & 0 \\ 0 & 196 & 0 & 1588 \end{bmatrix}.$$

Calculation of the inverse yields

$$(X'X)^{-1} = \begin{bmatrix} 1/3 & 0 & -1/21 & 0 \\ 0 & 397/1512 & 0 & -7/216 \\ -1/21 & 0 & 1/84 & 0 \\ 0 & -7/216 & 0 & 1/216 \end{bmatrix}.$$

The components of $\hat{\beta}$ are now obtained by post-multiplying this matrix by the matrix X' and the vector y. Such calculations yield

$$\hat{\beta}_1 = 38/7, \qquad \hat{\beta}_2 = 83/252, \qquad \hat{\beta}_3 = -1/28, \qquad \hat{\beta}_4 = 1/9.$$

Therefore the least squares third degree polynomial fitted to our data points is

$$y = 38/7 + (83/252)x - (1/28)x^2 + (1/9)x^3.$$

Since X_0 is obtained from X by deleting the last two columns of X, it follows that

$$X_0'X_0 = \begin{bmatrix} 7 & 0 \\ 0 & 28 \end{bmatrix}.$$

Hence,

$$(X_0'X_0)^{-1} = \begin{bmatrix} 1/7 & 0 \\ 0 & 1/28 \end{bmatrix}.$$

Further

$$X_0'y = \begin{bmatrix} 37 \\ 31 \end{bmatrix}.$$

Multiplying these two matrices will yield the desired estimates, namely,

$$\hat{\hat{\beta}}_1 = 37/7 \qquad \text{and} \qquad \hat{\hat{\beta}}_2 = 31/28.$$

As a result, the least squares polynomial of degree 1 fitted to our data points is the straight line

$$y = 37/7 + (31/28)x.$$

In order to apply (25), it is necessary to calculate the sum

$$\sum_{i=1}^{n} (y_i - y_i')^2,$$

both when y_i' represents the fitted third degree polynomial value and the first degree polynomial value corresponding to $x = x_i$. Calculations carried to two decimal places yielded the following sets of values for the y_i'.

Degree 3	1.12	3.74	4.95	5.43	5.83	6.83	9.10
Degree 1	1.97	3.07	4.18	5.29	6.39	7.50	8.61

Calculation of $\sum (y_i - y_i')^2$ is now readily carried out for both cases and will be found to give

$$(Y - X\hat{\beta})'(Y - X\hat{\beta}) = .34$$

and

$$(Y - X_0\hat{\beta})'(Y - X_0\hat{\beta}) = 3.12.$$

Application of (25) then gives

$$F = \frac{3.12 - .34}{.34} \frac{3}{2} = 12.$$

From Table VII in the appendix it will be found that the 5 percent right tail critical point corresponding to the degrees of freedom 2 and 3 is $F_0 = 9.55$. Since $12 > 9.55$, the hypothesis that a straight line will suffice as the regression curve will be rejected. A graph of the two fitted curves will show that the cubic fits the points very well, but that the straight line also does quite well. The superiority of the cubic over the linear function is strong enough to reject the linearity hypothesis, although barely so.

This problem will also be used to illustrate the application of Theorem 2 for finding confidence intervals for regression coefficients. We shall apply it by means of (21) to the problem of finding a 95 percent confidence interval for β_3. From the earlier calculation of $(X'X)^{-1}$, it will be seen that the value of σ_{33} is $1/84$. Since the calculations for that earlier problem also yielded the values $\hat{\beta}_3 = -(1/28)$ and $\sum (y_i - \hat{\mu}_i)^2 = .34$, and since the t_0 value for $n - k = 3$ degrees of freedom is 3.18, application of (21) will give

$$-\frac{1}{28} - 3.18 \sqrt{\frac{.34}{(84)(3)}} < \beta_3 < -\frac{1}{28} + 3.18 \sqrt{\frac{.34}{(84)(3)}}.$$

This simplifies to

$$-.153 < \beta_3 < .081.$$

If a confidence interval for a coefficient such as β_3 includes zero in its interior, then the test based on Theorem 2 of the hypothesis $H_0: \beta_3 = 0$ must accept H_0 because the acceptance region $|t| < t_0$ on which (21) is based is the noncritical region of the test based on Theorem 2. This result therefore includes the information that we would have accepted H_0

if we had tested it. This property of a confidence interval also serving as a test of a hypothetical value of the parameter is not restricted to the preceding type of problem. It is true in general if the noncritical region of the test is used to construct the corresponding confidence interval.

5.5. Analysis of variance

In Chapter 4 the analysis of variance model was introduced as a model for experimental design and was shown to be a special type of our linear regression model. Consider now the problem that was discussed there of deciding whether there are any appreciable differences between four brands of calculating machines in the time required to solve a set of problems.

If the same normality assumptions are made as before, the variables Y_{ij} will be independently normally distributed with a common variance σ^2 and with means given by the formula

(26) $\qquad \mu_{ij} = \mu + a_i + b_j, \qquad i = 1, \ldots, r, \qquad j = 1, \ldots, c.$

The problem of testing whether there are any essential differences in the brands of machines corresponds in this model to the problem of testing whether the column means are equal. Since the b's represent column differences with $\sum b_j = 0$, our problem can be formulated as the problem of testing the hypothesis

$$H_0 : b_1 = \cdots = b_c = 0.$$

In order to be able to apply Theorem 1 to this problem, it is necessary to find the least squares estimates $\hat{\mu}_{ij}$ and $\hat{\hat{\mu}}_{ij}$. They are obtained by means of (26) from the least squares estimates $\hat{\mu}, \hat{a}_i, \hat{b}_j$, and $\hat{\hat{\mu}}, \hat{\hat{a}}_i$, respectively. These first estimates were obtained in Chapter 4, Formulas (19) and (20), and were found to be

$$\hat{\mu} = \bar{y}, \qquad \hat{a}_i = \bar{y}_{i\cdot} - \bar{y}, \qquad \hat{b}_j = \bar{y}_{\cdot j} - \bar{y}.$$

Hence,

(27) $\qquad \hat{\mu}_{ij} = \hat{\mu} + \hat{a}_i + \hat{b}_j = \bar{y}_{i\cdot} + \bar{y}_{\cdot j} - \bar{y}.$

When H_0 is true the least squares estimates are those that minimize $\sum\sum (y_{ij} - \mu - a_i)^2$. It will be recalled from Chapter 4, however, that the calculation of the estimates for μ and a_i did not depend in any way upon the values of the b's; therefore, it follows directly from those same earlier results that

$$\hat{\hat{\mu}} = \bar{y} \qquad \text{and} \qquad \hat{\hat{a}}_i = \bar{y}_{i\cdot} - \bar{y}.$$

Hence,

(28) $\qquad \hat{\hat{\mu}}_{ij} = \bar{y}_{i\cdot} .$

The necessary sums of squares then assume the form

$$\sum_{i=1}^{r}\sum_{j=1}^{c}(y_{ij}-\hat{\mu}_{ij})^2 = \sum_{i=1}^{r}\sum_{j=1}^{c}(y_{ij}-\bar{y}_{i\cdot}-\bar{y}_{\cdot j}+\bar{y})^2$$

and

$$\sum_{i=1}^{r}\sum_{j=1}^{c}(y_{ij}-\hat{\mu}_{ij})^2 = \sum_{i=1}^{r}\sum_{j=1}^{c}(y_{ij}-\bar{y}_{i\cdot})^2.$$

As explained in Chapter 4, there are only $r + c - 1$ independent parameters in model (26); therefore we must express the vector β in terms of this many parameters if X is to be of full rank. A demonstration that X is of rank $r + c - 1$ is treated as an exercise at the end of this chapter. For convenience we shall choose β to be the vector whose transpose is $\beta' = (\mu, a_1, \ldots, a_{r-1}, b_1, \ldots, b_{c-1})$. Since $n = rc$, it therefore follows that $n - k = rc - (r + c - 1) = (r - 1)(c - 1)$.

In terms of our independent parameter choices, H_0 becomes the hypothesis $b_1 = \cdots = b_{c-1} = 0$. The matrix A that will make these restrictions a special case of $A\beta = 0$ is given by

$$A = \begin{bmatrix} 0 & 0\cdots0 & 1 & 0\cdots0 \\ 0 & 0\cdots0 & 0 & 1\cdots0 \\ \vdots & \vdots & \vdots & \vdots & \vdots & \vdots \\ 0 & 0\cdots0 & 0 & 0\cdots1 \end{bmatrix}$$

where the 1 occurring in the first row is in column $r + 1$ and there are $c - 1$ rows. It is obvious that A is of rank $c - 1$; therefore $q = c - 1$.

The substitution of the preceding results into (4) will produce the F variable

$$(29) \quad F = \frac{\sum_{i=1}^{r}\sum_{j=1}^{c}(y_{ij}-\bar{y}_{i\cdot})^2 - \sum_{i=1}^{r}\sum_{j=1}^{c}(y_{ij}-\bar{y}_{i\cdot}-\bar{y}_{\cdot j}+\bar{y})^2}{\sum_{i=1}^{r}\sum_{j=1}^{c}(y_{ij}-\bar{y}_{i\cdot}-\bar{y}_{\cdot j}+\bar{y})^2}$$

$$\times \frac{(r-1)(c-1)}{(c-1)}$$

with $c - 1$ and $(r - 1)(c - 1)$ degrees of freedom. If formula (19) is applied to this problem, it follows from (27) and (28) that F can be expressed in the following more compact form:

$$(30) \quad F = \frac{\sum_{i=1}^{r}\sum_{j=1}^{c}(\bar{y}_{\cdot j}-\bar{y})^2}{\sum_{i=1}^{r}\sum_{j=1}^{c}(y_{ij}-\bar{y}_{i\cdot}-\bar{y}_{\cdot j}+\bar{y})^2}(r-1).$$

The numerator in this F ratio is a measure of the variability of the column means, and therefore will tend to be excessively large when H_0 is false as compared to when H_0 is true. The denominator, however, is a measure of the variability that exists after the row and column effects have been subtracted, and therefore it will tend to have the same value whether H_0 is true or false. Thus, this ratio will tend to be excessively large when H_0 is false as compared to when H_0 is true, so it appears to be a good statistic for testing H_0. The justification for this test, of course, is that it is a likelihood ratio test and therefore possesses any optimal properties of such tests; however, it is satisfying to have a test agree with one's intuition.

In view of the reasonableness of F as a statistic for testing H_0, it would seem to be worth while to calculate the value of F and use that value as a basis for making a judgment concerning the truth of H_0, even though the assumptions that are needed to derive Theorem 1 are not satisfied as well as might be desired. In such situations, however, probability calculations based on the theorem cannot be relied upon. Some of the problems that arise when theorem assumptions are not satisfied will be discussed more fully in the next chapter.

As a numerical application of our analysis of variance F test, consider the problem of testing whether there are any differences in four brands of machines that were used by five workmen if the following data represent the units of production per day turned out by those workmen over an experimental period of time.

Machine

		1	2	3	4
	1	44	38	47	36
	2	46	40	52	43
Workman	3	34	36	44	32
	4	43	38	46	33
	5	38	42	49	49

Calculations here yielded the values

$$\sum_{i=1}^{5} \sum_{j=1}^{4} (\bar{y}_{\cdot j} - \bar{y})^2 = 338.8, \qquad \sum_{i=1}^{5} \sum_{j=1}^{4} (y_{ij} - \bar{y}_{i\cdot} - \bar{y}_{\cdot j} + \bar{y})^2 = 73.7.$$

Hence,

$$F = \frac{338.8}{73.7} (4) = 18.4.$$

From Table VII in the appendix, it will be found that the 5 percent critical point F_0 for 3 and 12 degrees of freedom is $F_0 = 3.49$; hence for $\alpha = .05$ the value of $F = 18.4$ certainly lies in the critical region, and H_0 will be rejected. The machines do differ in productivity as far as these workmen are concerned. Since we are rejecting H_0, we should return to the problem of estimating the differential effects of the four machines in order to determine the extent of the superiority of one or more machines over the others. The estimation problem was treated in the preceding chapter, so it is unnecessary to discuss it here. However, the proper procedure is to test first and then, if the test indicates there are real differences, to estimate.

By symmetry there exists a corresponding test for testing $H_0: a_1 = \cdots = a_r = 0$, given by replacing the numerator term of F in (30) by $\sum \sum (\bar{y}_{i\cdot} - \bar{y})^2 (c - 1)$. This would be used, for example, if we wished to test whether there are any significant differences in the productivity of the five workmen.

The analysis of variance model discussed here can be generalized to any number of control variables and to problems in which there is more than one observation in each cell. The addition of more control variables, however, requires that the model given by (26) be modified to incorporate the new parameters that arise. The extension to problems in which there is more than one observation in each cell is somewhat simpler, provided that each cell has the same number of observations. Modifications are necessary when the cell frequencies vary. In this connection, we shall carry out the derivation for the situation in which there is only one control variable that has been classified, but in which the number of observations for the different classifications varies. Thus, letting columns correspond to the various categories of the control variable, our model will assume the form

y_{11}		y_{1j}		y_{1c}
y_{21}		y_{2j}		y_{2c}
$\vdots$	$\cdots$	$\vdots$	$\cdots$	$\vdots$
$y_{n_1 1}$		$y_{n_j j}$		$y_{n_c c}$

Since we are assuming that the observed values in a given column are random sample values of a corresponding normal variable, we do not have row means here; therefore we must replace (26) by the model

$$\mu_{ij} = \mu + b_j, \qquad i = 1, \ldots, n_j, \qquad j = 1, \ldots, c.$$

As before, we wish to test the hypothesis

$$H_0: b_1 = \cdots = b_c = 0.$$

The sum of squares $\sum_{i=1}^{n} (y_i - \mu_i)^2$ now assumes the form

$$S = \sum_{j=1}^{c} \sum_{i=1}^{n_j} (y_{ij} - \mu_{ij})^2 = \sum_{j=1}^{c} \sum_{i=1}^{n_j} (y_{ij} - \mu - b_j)^2.$$

To obtain the minimizing values of μ and b_j, we proceed as we did in the preceding chapter when estimating analysis of variance parameters. First, we calculate

$$\frac{\partial S}{\partial \mu} = -2 \sum_{j=1}^{c} \sum_{i=1}^{n_j} (y_{ij} - \mu - b_j) = 0.$$

Upon summing the various terms, we obtain the equation

(31)
$$n\bar{y} - n\mu - \sum_{j=1}^{c} n_j b_j = 0,$$

where $n = \sum_{j=1}^{c} n_j$. Next, since $\sum_{j=1}^{c} b_j = 0$, we replace b_c by $-\sum_{j=1}^{c-1} b_j$ and write S in the form

$$S = \sum_{j=1}^{c-1} \sum_{i=1}^{n_j} (y_{ij} - \mu - b_j)^2 + \sum_{i=1}^{n_c} (y_{ic} - \mu - b_c)^2.$$

Then we calculate for $1 \le r \le c - 1$

$$\frac{\partial S}{\partial b_r} = -2 \sum_{i=1}^{n_r} (y_{ir} - \mu - b_r) + 2 \sum_{i=1}^{n_c} (y_{ic} - \mu - b_c) = 0.$$

Summation here will yield the equation

(32)
$$n_r \bar{y}_{\cdot r} - n_r \mu - n_r b_r = n_c \bar{y}_{\cdot c} - n_c \mu - n_c b_c.$$

This must hold for $r = 1, \ldots, c - 1$. It also obviously holds for $r = c$. Therefore, summing over those values of r, we will obtain

$$n\bar{y} - n\mu - \sum_{r=1}^{c} n_r b_r = r[n_c \bar{y}_{\cdot c} - n_c \mu - n_c b_c].$$

Because of (31) it follows that the quantity on the left, and hence the quantity in brackets on the right, must vanish, and therefore from (32) that we must have the solution

$$\hat{\mu} + \hat{b}_r = \bar{y}_{\cdot r}, \qquad r = 1, \ldots, c.$$

As a result

$$\sum_{j=1}^{c} \sum_{i=1}^{n_j} (y_{ij} - \hat{\mu}_{ij})^2 = \sum_{j=1}^{c} \sum_{i=1}^{n_j} (y_{ij} - \bar{y}_{\cdot j})^2.$$

When H_0 is assumed to be true, S assumes the form

$$S = \sum_{j=1}^{c} \sum_{i=1}^{n_j} (y_{ij} - \mu)^2.$$

It is obvious that the value of μ which minimizes this sum of squares is $\bar{y}$; hence

$$\sum_{j=1}^{c} \sum_{i=1}^{n_j} (y_{ij} - \hat{\mu}_{ij})^2 = \sum_{j=1}^{c} \sum_{i=1}^{n_j} (y_{ij} - \bar{y})^2.$$

By writing $(y_{ij} - \bar{y})^2 = [(y_{ij} - \bar{y}_{\cdot j}) + (\bar{y}_{\cdot j} - \bar{y})]^2$ and expanding the quantities in brackets as a binomial, it is readily verified that

$$\sum_{j=1}^{c} \sum_{i=1}^{n_j} (y_{ij} - \bar{y})^2 - \sum_{j=1}^{c} \sum_{i=1}^{n_j} (y_{ij} - \bar{y}_{\cdot j})^2 = \sum_{j=1}^{c} n_j (\bar{y}_{\cdot j} - \bar{y})^2.$$

Application of Theorem 1 then yields the desired F value, namely

$$F = \frac{\sum_{j=1}^{c} n_j(\bar{y}_{\cdot j} - \bar{y})^2}{\sum_{j=1}^{c} \sum_{i=1}^{n_j} (y_{ij} - \bar{y}_{\cdot j})^2} \frac{n-c}{c-1}.$$

The degrees of freedom here are $c - 1$ and $n - c$ because, as in our earlier model, there are only $c - 1$ independent parameters set equal to zero under H_0, and because there are c independent parameters in our model, namely, $\mu, b_1, \ldots, b_{c-1}$.

The preceding problem is essentially a generalization of the problem considered in Chapter 3 of testing whether the means of two independent normal variables are equal. For the case in which $c = 2$, the two tests are identical.

One of the striking advantages of the analysis of variance technique in designing experiments is that it enables the experimenter to study several control variables simultaneously and thereby discover in a relatively inexpensive manner which variables and which combination of variables are worth pursuing further. For example, suppose ten different seed varieties are tested for yield under varying growing conditions such as planting time, amount of fertilizer, and amount of water. If it is shown by the F test that they do differ significantly under some of the same growing conditions, then a more elaborate experiment can be designed to measure with greater precision the yield to be expected from, say, the top two or three producers under the desired growing conditions that produced maximum yield. Not only is such a design less costly than a design of consecutive individual experiments, but it also enables the experimenter to measure the effects of simultaneously varying his control variables at different levels.

There is an extensive literature on analysis of variance designs and their application to various fields. An understanding of the preceding theory of the general linear hypothesis should provide an excellent background for studying and understanding this important tool of experimental design.

Exercises

1 Given

$$f(y \mid \theta) = \frac{\exp\left[-(1/2) \sum_{i=1}^{n} (y_i - \beta_1 - \beta_2 x_{i2})^2\right]}{(2\pi)^{n/2}} \text{ and } H_0 \colon \beta_1 = 0,$$

write down the matrices X and A that will make this a special case of the general linear hypothesis model. What are the ranks of X and A? What assumptions are necessary concerning n and the x's to justify those ranks?

2 Given $f(y \mid \theta) = \dfrac{\exp\left[-(1/2) \sum_{i=1}^{n} (y_i - \mu_i)^2\right]}{(2\pi)^{n/2}}$, $\mu_i = \xi$, $i = 1, \ldots,$

m, $\mu_i = \eta$, $i = m + 1, \ldots, n$, and H_0: $\xi = \eta$, write down matrices X and A that will make this a special case of the general linear hypothesis model. What are the ranks of X and A?

3 Given $E(Y) = \beta_1 + \beta_2 x + \cdots + \beta_5 x^4$, H_0: $\beta_4 = \beta_5 = 0$, and the normality assumptions of Section 5.1, write down the matrices X and A that will make this a special case of the general linear hypothesis model.

4 In Exercise 3 assume that H_0 is the hypothesis H_0: $\beta_1 = \beta_2$, $\beta_3 = 0$, $\beta_4 = \beta_5$. Write down the matrix A that is needed for this hypothesis. What is the rank of A?

5 For the problem described in Exercise 3 carry out the algebraic steps needed to express μ in the form (8).

6 Given $f(y \mid \theta) = \dfrac{\exp\left[-(1/2) \sum_{i=1}^{3} (y_i - \mu_i)^2\right]}{(2\pi)^{3/2}}$ with $\mu_1 + \mu_2 + \mu_3 = 0$ and H_0: $\mu_1 = \mu_2$, show by a sketch in three dimensions the sample space and the subspaces $\mathscr{L}$ and $\mathscr{L}_0$, and the vectors corresponding to the sample point y, the mean μ, and the mean μ_0.

7 Given $f(y \mid \theta) = \dfrac{\exp\left[-(1/2) \sum_{i=1}^{n} (y_i - \mu_i)^2\right]}{(2\pi)^{n/2}}$ with $\mu_3 = \ldots = \mu_n = 0$ and H_0: $\mu_2 = 0$, describe the nature of the sample space and the subspaces $\mathscr{L}$ and $\mathscr{L}_0$.

8 Express the mean vector in Exercise 6 in the form $X\beta$ where β' is the row vector (μ_1, μ_2). Also express H_0 in the form $A\beta = 0$.

9 Express the mean vector in Exercise 7 in the form $X\beta$ where β' is the row vector (μ_1, μ_2). Also express H_0 in the form $A\beta = 0$.

10 Given $f(y_1, y_2) = \dfrac{\exp\left[-(1/2)[(y_1 - 2)^2 + (y_2 - 1)^2]\right]}{2\pi}$, find the density $g(z_1, z_2)$, if the basis vector α_1 makes a $45°$ angle with the y_1 axis and α_2 makes a $135°$ angle with the y_1 axis.

11 Show that the following transformation $z = Py$ is orthogonal. In which direction is α_1 oriented?

$$z_1 = \frac{1}{\sqrt{n}} y_1 + \cdots + \frac{1}{\sqrt{n}} y_n$$

$$z_2 = \frac{1}{\sqrt{2}} y_1 - \frac{1}{\sqrt{2}} y_2$$

$$z_3 = \frac{1}{\sqrt{6}} y_1 + \frac{1}{\sqrt{6}} y_2 - \frac{2}{\sqrt{6}} y_3$$

$$\vdots$$

$$z_n = \frac{1}{\sqrt{n(n-1)}} y_1 + \frac{1}{\sqrt{n(n-1)}} y_2 + \cdots - \frac{n-1}{\sqrt{n(n-1)}} y_n.$$

12 Use (24) to test the hypothesis $H_0 : b = 1$ for the linear regression model $E(Y \mid x) = a + bx$ of Exercise 3, Chapter 4. Use the calculations of that exercise and the formula given in Exercise 44 of Chapter 4 to assist you.

13 Use (24) to test the hypothesis $H_0 : b = 5$ for the linear regression model $E(Y \mid x) = a + bx$ of Exercise 2, Chapter 4. Use the calculations of that exercise to assist you.

14 Use (25) to test the hypothesis $H_0 : a = 0$ for the linear regression model $E(Y \mid x) = a + bx$ of Exercise 3, Chapter 4. Use the calculations of that exercise and those of Exercise 12 above to assist you.

15 For the problem that was used to illustrate the application of (25), test the hypothesis that the coefficient of x^2 is zero. The matrix $X_0'X_0$ is obtained from $X'X$ by deleting the third row and column of the latter. Begin at this stage, calculate $(X_0'X_0)^{-1}$, then calculate $(X_0'X_0)^{-1}X_0'y$.

16 Use (25) or (19) to test the hypothesis that the coefficient of x_2 is zero for the model of Exercise 20, Chapter 4. Use the calculations of that exercise and of Exercise 24, Chapter 4 to assist you.

17 Use (25) or (19) to test the hypothesis that the coefficients of both x_1 and x_2 are zero for the model of Exercise 20, Chapter 4. Use the calculations of Exercise 16 to assist you.

18 Use (25) or (19) to test the hypothesis $H_0 : b = 0$ for the parabolic regression model of Exercise 7, Chapter 4. Do not work this problem unless a calculating machine is available.

19 Show how to reformulate the linear regression model so that the hypothesis $H_0 : \beta_{l+1} = c_{l+1}, \ldots, \beta_k = c_k$ can be reduced to a hypothesis of the type $H_0 : \beta_{l+1} = 0, \ldots, \beta_k = 0$.

20 Use (21) and the computations that occurred in the solution of the illustrative exercise of Section 5.3 and those related to Figure 1 of Chapter 4 to find a 90 percent confidence interval for the coefficient b. Here $\sigma_{22} = 1/\sum (x_i - \bar{x})^2$.

21 Use (21) and the computations that occurred in the solution of the illustrative exercise of Section 5.4 to find a 95 percent confidence interval for the coefficient β_4. Comment on this interval and the interval found for β_3 in the text and their relationship to the testing problem discussed there.

22 Use (21) and the earlier computations that occurred in the solutions of Exercises 20 and 24, Chapter 4, to find 95 percent confidence intervals for the coefficients β_1 and β_2. Compare your results with the approximate intervals obtained for that exercise.

23 Write down the matrix X that will express μ as given by (26) in the form $X\beta$ where β' is the vector $(\mu, a_1, \ldots, a_{r-1}, b_1, \ldots, b_{c-1})$. Here

X must be a matrix with rc rows and $r + c - 1$ columns. Use the relations $a_r = -\sum_{i=1}^{r-1} a_i$ and $b_c = -\sum_{j=1}^{c-1} b_j$ in rows $c, 2c, \ldots,$ and in the last c rows.

24 Show that the matrix X in Exercise 23 is of rank $r + c - 1$, and hence that X is of full rank as required by Theorem 1.

25 For the illustrative exercise following (30), test the hypothesis that there are no differences among the five workmen as far as productivity is concerned.

26 Use (30) to test the hypothesis that there are no differences between (a) rations, (b) types for the experiment related to Exercise 31, Chapter 4.

27 Use (30) to test the hypothesis that there are no differences between (a) workmen, (b) machine types for the experiment related to Exercise 32, Chapter 4.

28 Show that the numerator of (29) reduces to the numerator of (30) by writing $[y_{ij} - \bar{y}_{i\cdot} - \bar{y}_{\cdot j} + \bar{y}]^2$ in the form $[(y_{ij} - \bar{y}_{i\cdot}) - (\bar{y}_{\cdot j} - \bar{y})]^2$, expanding it as a binomial, and then summing the three resulting terms in the proper orders.

29 Assume that $a_i = 0$, $i = 1, \ldots, r$, in the model of (26). This implies that it is known that there are no row differences and interest is centered on testing whether there are column differences. Apply Theorem 1 to this special case to obtain the F variable

$$F = \frac{\sum_{i=1}^{r} \sum_{j=1}^{c} (y_{\cdot j} - \bar{y})^2}{\sum_{i=1}^{r} \sum_{j=1}^{c} (y_{ij} - \bar{y}_{\cdot j})^2} \frac{c(r-1)}{c-1}.$$

30 Apply the formula of Exercise 29 to test the hypothesis that there are no differences in machine types for the experiment of Exercise 32, Chapter 4. Compare your F value with that obtained in Exercise 27 and explain any difference. Are both models appropriate here?

6 Nonparametric Methods

Most of the methods described in the preceding chapters require that the basic random variables possess certain familiar types of distributions. In particular, the last two chapters leaned heavily upon special normality assumptions. Although numerous theoretical and empirical investigations have shown that a number of those methods seem to be highly reliable even when the underlying distribution differs appreciably from the assumed distribution, this is not true for all of them.

When it is known that a basic variable does not possess the type of distribution required by our theory and when it is also known that the contemplated method is not reliable under modifications of that distribution, it is then necessary to find some alternative way for solving the problem. For continuous variable problems we sometimes have no idea of the nature of the distribution; therefore it would be desirable to have a method that is independent of it. Since our preceding methods have been concerned with the estimation or testing of the parameter θ for a density $f(x \mid \theta)$, a method that does not require the specification of $f(x \mid \theta)$ is usually called a *nonparametric method*.

When there is justification for assuming that f possesses a given functional form, it would be a serious mistake not to use that information. In constructing a statistical inference procedure, the more information that can be incorporated the more efficient the procedure is likely to become. Nonparametric methods should therefore be looked upon as alternatives to parametric methods when there is good reason to believe that the parametric method designed for the problem is not appropriate. There are, however, a number of nonparametric techniques that are nearly as good for a given problem as the best parametric method. Such a technique has several advantages over the parametric case. First, it has a wider range of applicability; and secondly, it does not require a careful check to see that the parametric assumption is justified. It is often impossible to prove or justify the parametric assumptions.

A number of nonparametric techniques have already been introduced in the preceding chapters, and some of them will be studied further in this chapter. In addition, a few new methods will be introduced to indicate how nonparametric methods are constructed.

6.1. Nonparametric estimation

A very general estimation problem that we can consider from a non-parametric point of view is that of estimating the distribution function $F(x)$ of a random variable X. This is certainly a nonparametric problem if no assumption is made concerning the nature of the unknown $F(x)$. As you will recall from Volume I, this function is defined by $F(x) = P(X \leq x)$. An estimate of $F(x)$, which is called the *empirical*, or *sample*, *distribution function* can be obtained in the following manner. The sample values of a random sample of size n are first arranged in order of size. Let $x_1 \leq x_2 \leq \cdots \leq x_n$ denote those ordered sample values. Then the empirical distribution function $S_n(x)$ is defined by means of the formula

(1)
$$ S_n(x) = \begin{cases} 0, & x < x_1 \\ \dfrac{i}{n}, & x_i \leq x < x_{i+1} \\ 1, & x_n \leq x \end{cases}. $$

For the purpose of illustrating how this estimate is constructed and to observe its accuracy, a sample of size 20 was taken of a normal variable with mean 2 and variance 1. The results of the sampling produced the following values, which have been ordered, correct to the nearest decimal: .3, .7, .9, 1.2, 1.4, 1.4, 1.5, 1.6, 1.9, 2.0, 2.1, 2.1, 2.3, 2.5, 2.6, 2.7, 3.0, 3.8, 3.9, 4.0. The empirical distribution function $S_{20}(x)$ is now easily obtained by means of (1). The theoretical distribution function of which $S_{20}(x)$ is an estimate is the function given by

$$ F(x) = \int_{-\infty}^{x} \frac{e^{-(t-2)^2/2}}{\sqrt{2\pi}} \, dt = \int_{-\infty}^{x-2} \frac{e^{-(s^2/2)}}{\sqrt{2\pi}} \, ds. $$

A graph of $F(x)$ is easily constructed by assigning convenient values to x and using Table IV in the appendix. The graphs of $F(x)$ and $S_{20}(x)$ are shown in Figure 1. It will be observed that $S_{20}(x)$ seems to approximate $F(x)$ well, except in the neighborhood of $x = 3.5$.

Since we are estimating an unknown distribution function $F(x)$, we will not be able to judge how good our estimate $S_n(x)$ is by a graphical comparison such as that given in Figure 1. A method for determining the accuracy of such an estimate is available, however, in the form of a confidence band for $F(x)$ that can be constructed with the aid of $S_n(x)$, as described below.

Assume that $F(x)$ is a known continuous function. Then it is possible to calculate the value of $|F(x) - S_n(x)|$ for any desired value of x. Furthermore, by studying this function it is possible to determine how large it can become as x ranges over its domain of values, because $F(x)$ is a continuous nondecreasing function satisfying $0 \leq F(x) \leq 1$, and $S_n(x)$ is a step

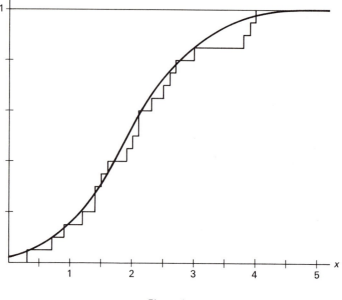

function with n steps. It is clear from Figure 1 that it therefore suffices to look at the left and right end points of an interval to determine how large the function $|F(x) - S_n(x)|$ can become in that interval. Since $S_n(x)$ is constant over each interval of the form $x_i \leq x < x_{i+1}$, it is possible for $|F(x) - S_n(x)|$ to assume its maximum value in such an interval at the left end point but not at the right end point because the right end point is not included in the interval. Thus, it is necessary to replace maximum by least upper bound in studying how large this function can become. We shall therefore be interested in the function

$$D_n = \sup_x |F(x) - S_n(x)|.$$

It now follows that it will suffice to evaluate $|F(x) - S_n(x)|$ at the end points of each interval, treating $S_n(x)$ as having a constant value in the closed interval $x_i \leq x \leq x_{i+1}$, and then choose the largest of those values for the value of D_n.

Since $S_n(x)$ is a function of our random sample, D_n is a random variable. A striking feature of this random variable is that its distribution, which we shall not derive, does not depend upon $F(x)$ as long as $F(x)$ is a continuous function; consequently, it can be used as the basis for a nonparametric method. We shall use it to construct a confidence band for $F(x)$.

The underlying reason why the distribution of D_n does not depend on the nature of F is that, as shown in Volume I, the random variable $F(X)$ possesses the uniform distribution over the interval $(0, 1)$, and that the value of D_n will be the same whether one works with X or $F(X)$. As a result one can just as well study the distribution of D_n in sampling from the uniform distribution as in sampling from the original distribution.

Assume that the distribution of D_n is available. Then it would be possible to find a value, which will be denoted by D_n^α, such that

$$P(D_n \le D_n^\alpha) = 1 - \alpha.$$

Such values have been calculated and are to be found in Table VIII of the appendix.

From the definition of D_n^α, it follows that

$$1 - \alpha = P\left(\sup_x |F(x) - S_n(x)| \le D_n^\alpha\right)$$

$$= P(|F(x) - S_n(x)| \le D_n^\alpha \text{ for all } x)$$

$$= P(S_n(x) - D_n^\alpha \le F(x) \le S_n(x) + D_n^\alpha \text{ for all } x).$$

The last equality shows that the probability is $1 - \alpha$ that the unknown distribution function $F(x)$ lies inside the band determined by the two step functions $S_n(x) - D_n^\alpha$ and $S_n(x) + D_n^\alpha$. The width of this confidence band serves as a measure of the accuracy of $S_n(x)$ as an estimate of $F(x)$.

To illustrate the technique for constructing a confidence band of this type, consider once more the data that gave rise to Figure 1. Suppose we desire a 90 percent confidence band. It will then be necessary to use $D_{20}^{.10}$ for the construction, which from Table VIII in the appendix will be found to be $D_{20}^{.10} = .26$. If this value is subtracted from and added to the step function sketched in Figure 1, the desired confidence band will be obtained. The result of this construction is shown in Figure 2. The normal distribution function $F(x)$ from which the samples were taken is represented by the smooth curve in that sketch. It lies well within the confidence band.

A confidence band such as this can also be used to test the hypothesis $H_0: F(x) = F_0(x)$, where $F_0(x)$ is some specified distribution function. If the graph of $F_0(x)$ lies inside the confidence band, H_0 is accepted, otherwise it is rejected. It is not necessary to graph $F_0(x)$, however, because its graph will lie inside the confidence band if, and only if, $D_n \le D_n^\alpha$. Thus, it suffices to calculate the value of D_n based on $F_0(x)$ and observe whether $D_n \le D_n^\alpha$. This use of D_n yields another method for solving the goodness of fit problem that was treated in Chapter 3 by means of the χ^2 test. The D_n statistic possesses the advantage that it is an exact method, whereas the χ^2 method requires a fairly large sample to justify the approximations that are needed in applying it.

As an illustration of how D_n may be applied to testing, consider the problem of testing the hypothesis that the data used to construct the empirical distribution of Figure 1 came from a normal distribution with mean 2 and variance 1. Since these data were obtained by sampling such a normal distribution, this is a true hypothesis. From Figure 1 it is clear that it suffices to calculate the value of $F(x_i)$, $i = 1, \ldots, n$, and compare it with the value of $S_n(x_i)$, or $S_n(x_{i-1})$, $i = 1, \ldots, n$, whichever produces the maximum absolute difference in determining the maximum difference in the graphs of $F(x)$ and $S_n(x)$. Here

$$F(x) = \int_{-\infty}^{x} \frac{e^{-(t-2)^2/2}}{\sqrt{2\pi}} \, dt = \int_{-\infty}^{x-2} \frac{e^{-(1/2)z^2}}{\sqrt{2\pi}} \, dz.$$

The values of $F(x_i)$ can be obtained from Table IV. These values, together with the associated values of $S_n(x_i)$, are the following:

x_i	.3	.7	.9	1.2	1.4	1.5	1.6	1.9	2.0
$F(x_i)$	.04	.10	.14	.21	.27	.31	.34	.46	.50
$S_n(x_i)$	.05	.10	.15	.20	.30	.35	.40	.45	.50

2.1	2.3	2.5	2.6	2.7	3.0	3.8	3.9	4.0
.54	.62	.69	.73	.76	.84	.96	.97	.98
.60	.65	.70	.75	.80	.85	.90	.95	1.00

The maximum values of $|F(x_i) - S_n(x_i)|$ and $|F(x_i) - S_n(x_{i-1})|$ are seen to be .06 and .11, respectively. The latter value occurs at $x = 3.8$. Thus, $D_n = .11$ here. Since the .05 critical value of D_n is given by $D_{20}^{.05} = .29$ and $D_n < D_n^\alpha$, the hypothesis is accepted.

The use of D_n for testing a hypothesis such as the preceding one is slightly out of place in this chapter, because specifying $F_0(x)$ usually requires specifying the parameters necessary to determine it, and then the test is not a nonparametric test. If we wished to test whether $F(x)$ is a distribution function of a certain family, such as a normal family, without specifying the parameters that determine a member of that family, it would be necessary to estimate those parameters from the sample and use them to obtain an estimate of $F(x)$. The random variable D_n based on this estimated $F(x)$ would, however, no longer possess the tabled distribution. Unfortunately, there is not a nice theory, such as was true for the χ^2 test, that makes the proper allowance for replacing unknown parameters by their sample estimates when applying this test.

A less general problem of nonparametric estimation is that of estimating properties of the distribution function. If it is assumed that the random

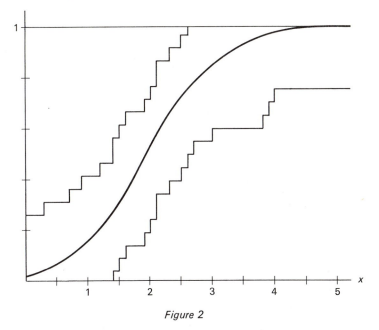

Figure 2

variable X possesses low order moments, then we might be interested in estimating those moments. Or, we might be interested in estimating certain points on the distribution, such as the median (midpoint), or the 5 percent right tail point. All of these are useful properties of the distribution and do not require that the distribution be specified by a density determined by parameters.

We have already studied the problem of estimating the moments of a random variable by means of the sample moments in Chapter 2. The question of the accuracy of those estimates is a difficult one and will not be treated here. We shall, however, consider the problem of estimating the median of a distribution in the next section under testing.

6.2. Nonparametric testing

Two of the most important parametric testing problems are those of testing a mean and testing the difference of two means. The corresponding nonparametric problems are those of testing a median and testing the difference of two medians. The median of a continuous random variable X is a point ξ such that $P(X < \xi) = P(X > \xi)$. If there is not an interval of values in the neighborhood of ξ where the density function is zero, this property will uniquely define the median. For a distribution that is symmetric about the median, the mean and median will be equal, provided the mean exists. Thus, the median should serve as a useful substitute for the mean in nonparametric situations.

In the following sections a few typical nonparametric tests will be presented without necessarily attempting to derive them or justify them. Most nonparametric tests have been obtained by analogy with corresponding parametric tests, because there is not a well developed theory of best tests for nonparametric problems. Thus, the purpose of this section is to present a few useful nonparametric tests and to indicate how such tests are constructed.

6.2.1. The sign test.   Assume that the continuous random variable X possesses the density function $f(x)$. Then the median, as defined before, is a point ξ such that

$$\int_{-\infty}^{\xi} f(x)\, dx = \int_{\xi}^{\infty} f(x)\, dx = \tfrac{1}{2}\,.$$

We shall assume that $f(x)$ is such that ξ is uniquely defined by this property. Now assume that a random sample of size n is to be taken of X and used to test the hypothesis

$$H_0: \xi = \xi_0 \qquad \text{against} \qquad H_1: \xi = \xi_1 > \xi_0.$$

Let the random sample be denoted by $X_1, \ldots, X_n$.   Then since $P(X > \xi_0 \mid H_0) = 1/2$, it follows that

$$P(X_i - \xi_0 > 0 \mid H_0) = 1/2, \qquad i = 1, \ldots, n.$$

Next, let

$$Z_i = \begin{cases} 0, & X_i - \xi_0 \leq 0 \\ 1, & X_i - \xi_0 > 0 \end{cases}, \qquad i = 1, \ldots, n.$$

The Z's are independent identically distributed random variables because the X's are such variables. Furthermore, when H_0 is true, $E(Z_i) = 1/2$; therefore when H_0 is true, the Z's represent a random sample of size n of a Bernoulli variable with $p = 1/2$. When H_1 is true, $E(Z_i) > 1/2$ because the frequency of positive values of $X_i - \xi_0$ will increase if the true median is larger than ξ_0. Hence, according to the methods of Chapter 3, we should choose the binomial variable $Z = \sum_{i=1}^{n} Z_i$ as our statistic for constructing the test, and we should use the right tail of the distribution of Z as critical region. Here

$$E(Z) = np = \frac{n}{2} \qquad \text{and} \qquad V(Z) = npq = \frac{n}{4}\,.$$

If n is sufficiently large to justify using the normal approximation to the binomial distribution, then the test should choose as critical region the region $\tau > \tau_0$, where $P(\tau > \tau_0) = \alpha$ and where

$$\tau = \frac{Z - (n/2)}{\sqrt{n/4}}\,.$$

Here τ is assumed to be a standard normal variable. If n is not sufficiently large to justify this approximation, it is necessary to find an integer z_0 satisfying

$$\sum_{z=z_0}^{n} \binom{n}{z} \left(\frac{1}{2}\right)^z \left(\frac{1}{2}\right)^{n-z} \le \alpha$$

and such that no smaller integer will satisfy the inequality. The critical region will then be the region where $Z \ge z_0$. As in most discrete variable problems, it may be necessary to introduce randomization at z_0, if a predetermined exact size α critical region is desired. Since the critical region does not depend upon the numerical value of ξ_1 but only on the fact that $\xi_1 > \xi_0$, our test applies equally well to the alternative $H_1 : \xi > \xi_0$.

As an illustration of how the sign test is applied, consider the data of Table 1 which lists the values of a random sample of size 20 of a normal variable with mean 0 and variance 4.

Table 1

1.90	2.50	.00	1.85	−.34	−2.61	.01	−.87	−2.14	.89
.70	1.17	−2.43	1.63	−3.25	.72	−2.25	.65	−.33	2.01

Let us test the hypothesis

$$H_0 : \xi = -1 \qquad \text{against} \qquad H_1 : \xi > -1.$$

We know that H_0 is false here, and therefore we should hope to reject it. First, we must calculate the values $x_i - \xi_0 = x_i + 1$. These are shown in Table 2.

Table 2

2.90	3.50	1.00	2.85	.66	−1.61	1.01	.13	−1.14	1.89
1.70	2.17	−1.43	2.63	−2.25	1.72	−1.25	1.65	.67	3.01

Since there are 15 positive values here, $Z = 15$. Using the normal approximation, we will obtain

$$\tau = \frac{15 - 10}{\sqrt{5}} = \sqrt{5} = 2.24.$$

For $\alpha = .05$ this value lies in the critical region $\tau \ge 1.64$; hence H_0 is rejected. When $p = 1/2$, the normal approximation to the binomial distribution is very good for n as large as 20. A more accurate approximation, however, for calculating right tail probabilities is obtained by replacing z by $z - 1/2$ in τ. This would have produced a value of $\tau = 2.01$, which still lies in the critical region.

Since our data were obtained from sampling a known normal variable, it is possible to solve this problem by parametric methods and then compare the two procedures. Although the value of σ is known to be 4 here, we shall assume that its value is not available and use the Student t test, which is known to possess certain optimal properties for this test when the normality assumption is justified.

The parametric problem is to test

$$H_0: \mu = -1 \qquad \text{against} \qquad H_1: \mu > -1.$$

Calculations with the data of Table 1 yield the values $\bar{x} = -.01$ and $s = 1.70$, and therefore the t value given by

$$t = \frac{\bar{x} - \mu}{s} \sqrt{n - 1} = \frac{-.01 + 1}{1.70} \sqrt{19} = 2.56.$$

Since the t variable is hardly distinguishable from a standard normal variable for a sample of this size, we may treat $t = 2.56$ as a standard normal variable value. It will be observed that the parametric test refutes the false hypothesis H_0 with more assurance than does the sign test, which produced a standard normal variable value that is somewhere between 2.01 and 2.24.

6.2.2. The rank-sum test.   In this section we shall consider the nonparametric analogue of testing the difference of two normal means. We shall formulate it as the problem of testing whether the continuous densities of two continuous random variables are the same, except possibly for the location of their medians. This problem would arise if, for example, a random variable X were sampled at two different times, and there was reason to believe that the median of the distribution had shifted to the right, or the left, during that period of time, but otherwise the distribution was unchanged. Let $f_1(x)$ and $f_2(x)$ denote the densities of the random variable X at the two sampling times, and consider the problem of testing

$$H_0: f_2(x) = f_1(x) \qquad \text{against} \qquad H_1: f_2(x) = f_1(x - c),$$

where c is a positive constant. Here H_1 indicates that $f_1(x)$ has shifted to the right by an amount c. The geometry of this relationship is shown in Figure 3.

Now assume that random samples $x'_1, \ldots, x'_{n_1}$ and $y'_1, \ldots, y'_{n_2}$ have been taken from $f_1(x)$ and $f_2(x)$, respectively. Let these two sets of values be arranged in order of magnitude and denote the ordered values by $x_1, \ldots, x_{n_1}$ and $y_1, \ldots, y_{n_2}$, respectively. Next, let these ordered sets be combined into a single ordered set. It might be, for example, the set of values

$$x_1, x_2, y_1, x_3, y_2, x_4, x_5, y_3, \ldots, y_{n_2}.$$

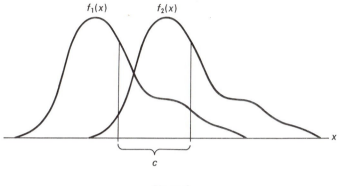

$f_1(x)$ $f_2(x)$

x

c

Figure 3

The test that is about to be described is based on the *randomization principle*. This principle states that tests shall be based on the distribution of a statistic under all possible permutations of the sample values. In standard parametric tests the distribution of a statistic is that which would arise under repeated random sampling of $f(x)$, whereas here there are no new samples contemplated. Each permutation of the given sample values plays a role here similar to a fresh random sample for a parametric test.

When H_0 is true, the preceding combined ordered values represent the ordered values of a random sample of size $n_1 + n_2$ taken from $f_1(x)$. Under random sampling every possible order in which these sample values could have been obtained should have the same probability of occurrence. Thus, if a random sample of size three yielded the values 8, 10, 14 in some order unknown to us, then all six possible permutations of those three numbers should be assigned the same probability of being the correct, but unknown, order. Under this assumption of equal probability for all orders in our combined set, each such order will be assigned the probability

$$\frac{1}{(n_1 + n_2)!}.$$

To construct a test based on the randomization principle, it is necessary to choose a statistic that will be effective in discriminating between H_0 and H_1. Since the y values will tend to be larger than the x values when H_1 is true, our statistic should be one that capitalizes on this fact. One such statistic that has proved very effective in doing this is the sum of the ranks of the y's in the combined ordered set. Thus, in the preceding illustration of a typical ordering, we would first write down the numbers 3, 5, 8, ..., $n_1 + n_2$, which are the ranks of $y_1, y_2, y_3, \ldots, y_{n_2}$, and then sum those numbers. If $r(y_i)$ denotes the rank of y_i in the combined ordered set, the statistic that will be used for constructing our test is

$$T = \sum_{i=1}^{n_2} r(y_i).$$

Since H_1 will tend to favor large values of T relative to those that will be obtained when H_0 is true, the critical region should consist of the right tail of the distribution of T values when H_0 is true. Now the randomization principle requires us to find the distribution of T under all possible permutations of the sample values. Since the ranking of values does not depend upon the magnitudes of the sample values but only upon their relative magnitudes, it suffices to look only at the positions of the y's in the combined ranking in order to evaluate T. Permuting the x's in their n_1 positions and the y's in their n_2 positions will not affect the value of T. Since there are $n_1! \, n_2!$ ways of permuting the x's and the y's in their positions, the probability that should be attached to each distinguishable combined ordering will be $\dfrac{n_1! \, n_2!}{(n_1 + n_2)!}$.

The problem now reduces to finding the number of distinguishable orderings that will produce the same value of T and do this for all possible T values. If the T values are arranged in order of size and the proper probabilities are attached to them, the desired distribution of T under randomization will have been obtained. The probability to be attached to any T value will be $\dfrac{k_T n_1! \, n_2!}{(n_1 + n_2)!}$, where k_T is the number of distinguishable orderings that produced that value of T. If this distribution were available, it would be possible to find a value T_0 such that $P(T \geq T_0) \leq \alpha$.

Since T values depend only upon relative ranks and not on the magnitudes of the sample values, the problem of finding the distribution of T is one of combinatorics. It is clear that the computations required to obtain the distribution of T would be exceedingly lengthy even for moderate values of n_1 and n_2. Fortunately, such computations have been carried out for n_1 and n_2 not larger than 10 each. Those computations have been used to find useful left tail and right tail points of the distribution. In particular, the integer values of T, call them T_1 and T_2, that came closest to satisfying $P(T \leq T_1) = \alpha = P(T \geq T_2)$ for $\alpha = .025$ and $\alpha = .05$ were calculated. These values have been listed in Table IX in the appendix for various combinations of values of n_1 and n_2, where for convenience $n_1 \leq n_2$. Thus, the larger of the two samples is associated with the y's. If the slippage of $f_1(x)$ is to the left, then of course the left tail of the T distribution will contain the critical region, which means that T_1 will serve as the critical point.

For values of n_1 and n_2 larger than the tabled values, there exists a satisfactory normal curve approximation to the distribution of T. It can be shown that when H_0 is true, $n_1 \geq 10$ and $n_2 \geq 10$, T will possess an approximate normal distribution with mean and variance given by the formulas

$$E(T) = \frac{n_1(n_1 + n_2 + 1)}{2}$$

$$V(T) = \frac{n_1 n_2(n_1 + n_2 + 1)}{12}.$$

To illustrate the use of this test, the sample values in the second row of Table 1 were altered by adding 1 to each of them. Thus, H_1 is true here with $c = 1$. After this modification the sample values in both rows were ordered and give the values listed in Table 3.

Table 3

x_i	-2.61	-2.14	$-.87$	$-.34$	.00	.01	.89	1.85	1.90	2.50
y_i	-2.25	-1.43	-1.25	.67	1.65	1.70	1.72	2.17	2.63	3.01

The combined ordering of these values, in which the y values have been underlined for easy recognition, then becomes

$$-2.61, \quad -\underline{2.25}, \quad -2.14, \quad -\underline{1.43}, \quad -\underline{1.25}, \quad -.87, \quad -.34, \quad .00, \quad .01, \quad \underline{.67},$$

$$.89, \quad \underline{1.65}, \quad \underline{1.70}, \quad \underline{1.72}, \quad 1.85, \quad 1.90, \quad \underline{2.17}, \quad 2.50, \quad \underline{2.63}, \quad \underline{3.01}.$$

The ranks of the y's in this ordered set are 2, 4, 5, 10, 12, 13, 14, 17, 19, 20; hence the value of T is 116. From Table IX in the appendix it will be found that the .05 right tail critical point for $n_1 = n_2 = 10$ is 127; therefore H_0 will be accepted. This test was not able to discover the fact that the median had been shifted one unit to the right for the second sampling experiment.

The standard parametric test to employ here, if we assume normal distributions with a common unknown variance, is the Student t test. It should be interesting to determine whether this test will discover the shift in the mean. We shall ignore the fact that the variance is actually known. Calculations based on the Student t variable of Chapter 3 will yield $t = .93$ with 18 degrees of freedom. Thus, this test also fails to reject H_0. Since n_1 and n_2 are sufficiently large to justify using the normal approximation to T, let us solve the problem by the large sample method. Calculations here yield $E(T) = 105$, $V(T) = 175$. For $T = 116$ we therefore obtain

$$\tau = \frac{116 - 105}{\sqrt{175}} = .83.$$

Since the t distribution for 18 degrees of freedom is practically equivalent to that of a standard normal variable, the value $t = .93$ is not much better than the value $\tau = .83$ in trying to discover the shift that occurred. The rank-sum test is known to be an excellent test for detecting slippage even when the normality assumption is satisfied, so that it should come as no surprise to find so little difference between these two tests for this problem. The rank-sum test is sometimes called the Wilcoxon-Mann-Whitney test because all three individuals were involved in some phase of its development.

6.3. Randomness

All the tests that have been constructed in the preceding chapters have been based on random samples. Such samples require that the random variables $X_1, \ldots, X_n$ be independent variables. For a variable such as the maximum temperature during the day, it is clear that a set of consecutive daily temperatures can hardly be treated as a set of independent variables. This will be true of many other variables for which measurements are taken over time. Since the conclusions based on tests that assume random sampling are not likely to be valid when this assumption is not justified, it is essential to have some method for determining whether we are justified in believing that we are taking random samples.

For data such as consecutive daily temperatures, it is clear that the observations X_i and X_{i+1} will tend to be positively correlated; therefore a test that capitalizes on this correlation would be effective in discovering this type of dependence. If our data had consisted of the consecutive daily number of automobile accidents in a given city, we might have expected the observations X_i and X_{i+7} to be correlated because the amount of driving during the various days of the week varies considerably. This type of departure from randomness could probably be discovered by means of the correlation between values seven days apart. More generally, for data that are cyclical in nature, a correlation coefficient based on a time lag corresponding to the distance between consecutive peaks in the sequence of values should prove effective in demonstrating a lack of randomness.

We cannot use the ordinary sample correlation coefficient to construct a test of randomness because its distribution requires a parametric distribution assumption on X and Y; therefore we shall resort to the nonparametric randomization device that was introduced in the preceding section. If we choose the correlation coefficient r as our statistic, this implies that we must calculate the value of r for every possible permutation of our data, and from such calculations obtain the distribution of r, given the sample values. Large positive or negative values of r would naturally be placed in the critical region of the test.

If there is reason to believe that neighboring pairs of observations are correlated, we would calculate the correlation coefficient based on choosing $Y_i = X_{i+1}$. This means calculating r for the following pairs of numbers

x_i	x_1	x_2	$\cdots$	x_i	$\cdots$	x_{n-1}
y_i	x_2	x_3	$\cdots$	x_{i+1}	$\cdots$	x_n

The theory becomes somewhat simpler if an extra pair of numbers (x_n, y_n) is added to the end of this set, so that there will be n pairs of values. This will be done by defining $y_n = x_{n+1} = x_1$. The resulting correlation co-efficient is called the *circular form of the serial correlation coefficient.* For large values of n this modification will have no appreciable effect on the value of r unless the x's were climbing or falling rapidly so that there would be a large difference between the first and last values. Without this modification r is called the *serial correlation coefficient with lag* 1.

A standard calculating formula for the sample correlation coefficient is the following one:

$$r = \frac{\sum_{i=1}^{n} x_i y_i - n\bar{x}\bar{y}}{n s_x s_y}.$$

In order to obtain the circular form of the serial correlation coefficient with lag 1, it is necessary to substitute x_{i+1} for y_i and x_1 for x_{n+1} in this formula. Now the sample mean $\bar{x}$ and the sample standard deviation s_x of a set of observational values are independent of the order in which those observations were obtained; hence those quantities do not change under permutations of the sample values $x_1, \ldots, x_n$. This is also true for $\bar{y}$ and s_y, because with $y_n = x_1$ the same set of values are involved as for $\bar{x}$ and s_x. Thus, the only part of r that changes under permutations is the term $\sum_{i=1}^{n} x_i x_{i+1}$. Letting

$$R = \sum_{i=1}^{n} x_i x_{i+1},$$

it therefore suffices to find the distribution of R under all possible permutations of the values $x_1, \ldots, x_n$.

The randomization principle as it is applied to R differs from that applied to T of the preceding section in that here the value of R does depend upon the magnitudes of the x's and not merely upon their relative ranks. As a consequence it is not possible to construct a table of critical values for the distribution of R by combinatorial methods as was the case for the distribution of T. Here it is necessary to consider only problems for which n is so large that an asymptotic result can be applied. If it is assumed that the sample values $x_1, \ldots, x_n$ satisfy some modest regularity conditions, which will usually be satisfied if they represent a random sample from any reasonable distribution, then R will possess an asymptotic normal distribution with mean and variance given by the formulas

$$E(R) = \frac{S_1^2 - S_2}{n - 1},$$

$$V(R) = \frac{S_2^2 - S_4}{n - 1} + \frac{S_1^4 - 4S_1^2 S_2 + 4S_1 S_3 + S_2^2 - 2S_4}{(n - 1)(n - 2)} - E^2(R),$$

where $S_k = x_1^k + \cdots + x_n^k$.

The preceding theory is also applicable to R if it represents the correlation between pairs of values that are lagged more than one unit apart; therefore it is easily adapted to the problem of testing whether there exists a cycle of fixed length in a sequence of values. It has been assumed throughout this discussion that we are using the circular form of the serial correlation coefficient. For a lag of length k it is therefore necessary to use the first k values of x for the last k values of y in computing R.

Since the correlation coefficient is unaffected by adding a constant to each x value or multiplying each x value by the same constant, the test based on R must also be independent of such operations. As a result it may be possible to simplify the computations needed to evaluate R somewhat by employing such techniques.

As an illustration of how this test is applied, consider the following sequence of numbers obtained from a table of two-digit random numbers, but with only numbers in the interval $[40, 60]$ being retained: 45, 41, 50, 41, 59, 57, 56, 47, 40, 42, 54, 44, 49, 52, 53. It will be convenient to subtract 50 from each value to give the sequence $-5, -9, 0, -9, 9, 7, 6, -3, -10,$ $-8, 4, -6, -1, 2, 3$. Calculations will then give

$$S_1 = -20, \quad S_2 = 592, \quad S_3 = -1,952, \quad S_4 = 39,832.$$

Additional calculations will yield the values $E(R) = 13.7$, $V(R) = 20,022$, and $R = 100$. Hence

$$\tau = \frac{R - E(R)}{\sqrt{V(R)}} = \frac{86.3}{141.5} = .61.$$

This value is well within the acceptance region for a standard normal variable test; therefore there is no reason for rejecting the hypothesis that the sample values behave like a random sample as far as correlation between neighboring pairs is concerned. Since we obtained our sample from a table of random numbers, we would have been surprised if H_0 had not been accepted.

The calculations needed to carry out this test are rather heavy unless n is small or unless fast computing facilities are available. There are numerous other nonparametric tests available for testing randomness that are considerably easier to calculate. This particular test was chosen, however, to illustrate a test based on the randomization principle as applied to sample values rather than to sample ranks, and because serial correlation is a very useful concept in the study of stochastic processes, which will occur in Volume III.

6.4. Robustness

It is to be expected that a better statistical method can be constructed if useful assumptions are permitted than if this is not the case; therefore we should expect to find that a good parametric method for solving a problem is superior to the corresponding nonparametric one. For example, we should expect the test based on the *t* statistic for testing a normal mean to be superior to the sign test for testing a median, provided we know that the normality assumption is justified. Similarly we should expect the *t* test for testing the equality of two normal means to be superior to the rank-sum test for that same problem when the necessary assumptions are satisfied. A numerical comparison of those two methods on a pair of random samples was made in Section 6.2.2; and for the two problems considered there, it appeared that there was a slight advantage for the parametric technique, particularly when testing a single mean.

In view of the general superiority of good parametric over corresponding nonparametric methods, it is important to know how valid a parametric method is under modifications of its basic assumptions. If a test is not appreciably affected by moderate modifications, there is little point in switching to a nonparametric method unless the assumptions are seriously violated. A test that is reliable under rather strong modifications of the assumptions on which it was based is said to be *robust*. If a good robust test is available for a problem, then there is little need for a nonparametric technique to solve it.

The assumption of normality that occurred in so many of the techniques developed in the preceding chapters is a severe restriction. We can not in practice expect most variables to possess this type of distribution. However, when the statistic that is used to construct a test is a sample mean and the sample size is not too small, the Central Limit Theorem assures us that the sample mean will be approximately normally distributed even though the basic random variable has a distribution far removed from a normal one. As a result, many tests based on sample means are quite robust, and therefore the normality assumption is not as restrictive as it might seem to be. In this connection, the *t* test for testing the difference of two means requires the additional assumption that the variances of the two variables being sampled be equal. Large differences in the variances can be harmful and thereby destroy the robustness of that test. However, the rank-sum test that was introduced as a competitor is also harmed by such large differences. The test using a chi-square variable to test the hypothesis $H_0: \sigma = \sigma_0$, which was derived on the basis of a normality assumption, is an example of a test that lacks robustness with respect to that assumption.

For the purpose of illustrating whether a statistical procedure has robustness or not, we shall choose the non-normal density $f(x) = xe^{-x}$, $x > 0$. We shall study what effect this lack of normality will have on the standard procedures for testing whether the mean and variance of the random variable X possess postulated values μ_0 and σ_0^2.

We know that $E\overline{X} = \mu$ and $V(\overline{X}) = \sigma^2/n$, regardless of the distribution of X as long as X possesses a second moment, which it obviously does here. For simplicity of explanation we shall assume that n is so large that we may safely replace σ^2 by s^2, and thereby treat σ^2 as known when testing a hypothetical value of μ. Then the large sample technique for testing $H_0: \mu = \mu_0$ against $H_1: \mu \neq \mu_0$, based on assuming that X is a normal variable, chooses as acceptance region the interval given by

$$|\bar{x} - \mu_0| < \tau_0 \frac{s}{\sqrt{n}},$$

where τ_0 is the proper standard normal variable value corresponding to the selected value of α. Now from the Central Limit Theorem we know that $\overline{X}$ will possess an approximate normal distribution with these same parameter values for large n, if X possesses the density $f(x) = xe^{-x}$, $x > 0$; therefore this test is appropriate in either case if n is large. Empirical investigations will also show that for our choice of non-normal density, n does not need to be very large for this robustness property to hold. Although we assumed that n was sufficiently large to permit us to replace σ by s in the calculations, empirical investigations have shown that the test based on Student's t statistic is also robust. This also applies to testing the difference of two means when the two variances are equal.

Next, consider the corresponding comparison when a test of $H_0: \sigma^2 = \sigma_0^2$ against $H_1: \sigma^2 \neq \sigma_0^2$ is to be made by large sample methods. Under a normality assumption on X it is customary to choose as acceptance region the region given by

$$\chi_1^2 < \frac{ns^2}{\sigma^2} < \chi_2^2.$$

However, we shall assume that n is so large that we may apply the Central Limit Theorem to nS^2/σ^2 and thereby treat it as an approximate normal variable. Although only simple sums were considered in this connection in Volume I, the Central Limit Theorem is of wide applicability and may be applied to the type of sum occurring in nS^2. Thus, we may treat $w = nS^2/\sigma^2$ as an approximate normal variable.

Under the assumption that X is a normal variable, we know that nS^2/σ^2 possesses a chi-square distribution with $n - 1$ degrees of freedom. Furthermore, from the properties of a chi-square variable we know that

$$E(w) = n - 1 \qquad \text{and} \qquad V(w) = 2(n - 1).$$

Using these values and treating nS^2/σ^2 as an approximate normal variable, the acceptance region for our large sample test is given by

(2)
$$\left| \frac{ns^2}{\sigma_0^2} - (n-1) \right| < \tau_0 \sqrt{2(n-1)}.$$

Now consider what modifications occur in this test when X possesses the density $f(x) = xe^{-x}$, $x > 0$. Tedious computations with the expected value operator E will show that the following formula is valid for any random variable that possesses a fourth moment.

$$V(S^2) = \frac{\mu_4 - \mu_2^2}{n} - \frac{2(\mu_4 - 2\mu_2^2)}{n^2} + \frac{\mu_4 - 3\mu_2^2}{n^3}.$$

The μ's here represent moments about the mean; therefore $\mu_2 = \sigma^2$. As a result

$$V\left(\frac{nS^2}{\sigma^2}\right) = \frac{n^2}{\mu_2^2}\left[\frac{\mu_4 - \mu_2^2}{n} - \frac{2(\mu_4 - 2\mu_2^2)}{n^2} + \frac{\mu_4 - 3\mu_2^2}{n^3}\right].$$

Since we are choosing n to be very large, we will ignore the last two terms; therefore we may write

$$V\left(\frac{nS^2}{\sigma^2}\right) = n\frac{\mu_4 - \mu_2^2}{\mu_2^2}.$$

For the density $f(x) = xe^{-x}$, $x > 0$, calculations give the values $\mu_2 = 2$ and $\mu_4 = 24$; consequently for X possessing this density

$$V\left(\frac{nS^2}{\sigma^2}\right) = 5n.$$

We know from earlier work that for any random variable possessing a second moment

$$E(S^2) = \frac{n-1}{n}\sigma^2,$$

and therefore that

$$E\left(\frac{nS^2}{\sigma^2}\right) = n - 1.$$

The large sample test based on treating nS^2/σ^2 as an approximate normal variable then chooses as acceptance region the region given by the inequality

(3)
$$\left| \frac{ns^2}{\sigma_0^2} - (n-1) \right| < \tau_0 \sqrt{5n}.$$

A comparison of this large sample test with the earlier large sample test (2) based on a normality assumption shows that a serious error would be introduced if normality were assumed when in fact X possesses the density $f(x) = xe^{-x}$, $x > 0$. The correct test given by (3) uses an acceptance interval that is approximately $\sqrt{5/2} = 1.58$ times as long as the corresponding interval given by assuming normality. As a result the hypothesis H_0 will be rejected much too frequently here if the test based on normality is used. This example illustrates a test that does not possess robustness with respect to the normality assumption. Although the assumptions and calculations here were a bit crude, it is hoped that these two illustrations will indicate how some tests are robust and others are not.

In view of the lack of robustness of the classical test for σ^2, it is clear that one should not use that test unless it is known that the variable X possesses a normal distribution to a very good approximation. If such information is not available, a nonparametric test, or else a more robust test, should be employed. For large samples a test of the latter type is readily constructed by using the preceding calculations and Central Limit Theorem arguments in the following manner.

Since the earlier discussion indicated that the variable $w = nS^2/\sigma^2$ may be treated as an approximate normal variable with $E(w) = n - 1$ and $V(w) = n(\mu_4 - \mu_2^2)/\mu_2^2$ when n is large, it follows that

$$(4) \qquad \tau = \frac{(nS^2/\sigma^2) - (n - 1)}{\sqrt{n(\mu_4 - \mu_2^2)/\mu_2^2}} = \frac{\sqrt{n}(S^2 - ((n - 1)/n)\sigma^2)}{\sqrt{\mu_4 - \mu_2^2}}$$

may then be treated as an approximate standard normal variable. This will also be true if μ_4 and μ_2^2 are replaced by their sample moment estimates m_4 and m_2^2. Thus, the variable

$$z = \frac{\sqrt{n}(m_2 - ((n - 1)/n)\sigma^2)}{\sqrt{m_4 - m_2^2}}$$

will behave like a standard normal variable for large n, and may therefore be used to test hypotheses about σ^2 or to find confidence intervals for σ^2, when one is not justified in assuming that X possesses an approximate normal distribution, but is justified in assuming that X possesses low order moments.

The normality assumption that was introduced in Chapter 5 to yield the F test for the general linear hypothesis is also not as stringent a restriction as might be expected. Both theoretical and empirical investigations have shown, for example, that the 5 percent critical point of the F variable is not altered appreciably if the basic variable has a distribution that differs considerably from normality. Thus, the F test is a robust test with respect to the normality assumption. These same investigations have shown, however, that if the value of σ^2 varies considerably in going from column to column in the analysis of variance test, then the 5 percent critical value will be altered appreciably. If the largest value of σ^2 is three times the smallest value, the .05 critical value of F listed in Table VII corresponding to small degrees of freedom may occur at the .07 right tail value of the true distribution of F. Thus, there will be a tendency to reject H_0 more frequently than is justified at the .05 level. If the largest value of σ^2 is ten times the smallest value, the listed .05 critical value of F may occur as far away as at the .12 value of the true distribution of F. Such large variation in the value of σ^2 is therefore a serious matter when it comes to applying the F test to analysis of variance problems. The F test is therefore not a robust test with respect to large variation in the value of σ^2, but it is with respect to the normality assumption.

In some regression problems there will be considerable variation in the variance over the range of x values. Thus, there may be much more variability near the middle of the interval of values than near the extremities; however, the variability is seldom sufficiently severe to harm the F test appreciably. A more serious matter, however, in many regression studies is the lack of independence of the variables, particularly when the x's represent times at which observations are taken. Nonparametric methods will be of no assistance in such situations, however, because independence is required for both methods.

The nonparametric methods that have been discussed in this chapter are but a small sample of the many such methods available. The few that have been presented represent some of the more useful techniques for solving those problems. The literature on nonparametric statistics is a very extensive one. The parametric methods that have been presented in this book also represent a small but important segment of available methods. The field of mathematical statistics is a very large one and this book has merely attempted to give the student an insight into the nature of that field.

Exercises

1 Take a random sample of size 20 from the uniform density $f(x) = 1$, $0 \le x \le 1$, by choosing 20 random two-digit numbers from Table II in the appendix and dividing them by 100. Use these approximate random sample values from $f(x)$ to construct a 90 percent confidence band for $F(x)$. Graph $F(x)$ to see whether it lies inside the band.

2 The following set of numbers is from the table of random normal numbers with $\mu = 0$ and $\sigma = 1$ found in Table III in the appendix. Use them to construct an 80 percent confidence band for $F(x)$. Graph $F(x)$ to see whether it lies inside this band.

$$-1.80, \ -2.01, \quad .54, \ -1.63, \quad .25, \ -.17, \quad .03, \ .08, \quad .47, \ -1.03,$$
$$.20, \quad .21, \ -1.08, \quad -.22, \ -.29, \quad 1.22, \ 1.12, \ .00, \ 2.01, \ -.59.$$

3 Use the statistic D_n to test the hypothesis that $f(x) = 1$, $0 \le x \le 1$, (which is true here) for the data obtained in Exercise 1. Choose $\alpha = .05$.

4 Use the statistic D_n to test the hypothesis that $f(x) = \dfrac{e^{-x^2/2}}{\sqrt{2\pi}}$ (which is true here) for the data of Exercise 2. Choose $\alpha = .05$.

5 Use the statistic D_n to test the hypothesis that the following data are a random sample from a normal density with $\mu = 36$ and $\sigma = 2$. Choose $\alpha = .20$.

32.4, 34.4, 32.0, 34.5, 34.3, 34.7, 38.0, 35.6, 37.2, 31.6.

6 Toss six coins thirty times and record the number of heads obtained in each of the thirty experiments. Use the statistic D_n to test the hypothesis that the data are from a binomial density with $n = 6$ and $p = 1/2$ (which is true here). For a discrete variable such as this, the statistic D_n tends to be too conservative in the sense that the value of α is usually smaller than the value given by the table of values of D_n^α which were obtained for a continuous $F(x)$. Work this same problem by the large sample parametric method.

7 Eighty rats were separated into forty pairs by attempting to form pairs that were similar in size and appearance. One member of each pair was fed diet A and the other member diet B. At the conclusion of the experiment it was found that 28 of the diet A rats outgained their diet B companions. Would you reject the hypothesis of no difference on this evidence using the sign test?

8 The following data represent the resistance per unit length of wire tested from consecutive shipments of wire. Use the sign test to test the hypothesis that the median of the distribution is .50. Discard .50 values in applying the test.

.51, .49, .48, .48 .47, .51, .50, .52, .46, .47, .48, .53, .47, .51, .49, .49, .52, .46, .44, .54, .50, .49, .46, .51, .54, .46, .47, .52, .48, .44.

9 Use the data of Exercise 2 to test the hypothesis $H_0 : \mu = 1/4$ (which is false here) against $H_1 : \mu \neq 1/4$ by means of the sign test. Compare the sign test here with the test based on the statistic $\overline{X}$ under the normality assumption $f(x \mid \mu) = \dfrac{\exp\left[-(x - u)^2/2\right]}{\sqrt{2\pi}}$.

10 In a reading experiment in an elementary school, sixteen pairs of children were chosen with each pair matched with respect to ability and background. One child was taught to read by method A and the other child by method B. The results of a test at the end of the experimental period are listed below. Use the sign test to test the hypothesis of no difference in the two methods.

A	66	69	70	62	64	62	72	76	78	64	73	80	67	74
B	64	68	69	60	66	61	70	75	72	65	70	78	68	72

11 Choose 30 two-digit random numbers from Table II in the appendix. Let the first 10 represent X values and the last 20 represent Y values. Apply the rank-sum test to test the hypothesis that X and Y possess the same distribution (which is true here).

12 The following data are random sample values from a non-normal distribution taken at two different times. There is reason to believe that the median of the distribution may have shifted to the right between the sampling periods. Do these data justify that belief on the basis of the rank-sum test?

I	72, 21, 32, 52, 73, 40, 31, 62, 40, 42
II	24, 56, 50, 30, 53, 78, 33, 70, 15, 29, 36, 60, 57, 79, 51

13 Take random samples of size 10 each from the two uniform densities given by $f_1(x) = 1, 0 \le x \le 1$, and $f_2(x) = 2, 0 \le x \le 1/2$, by choosing two-digit random numbers from Table II in the appendix, dividing these numbers by 100 and discarding any number above .50 when sampling $f_2(x)$. Use these approximate sample values and the rank-sum test to test the hypothesis that $f_1(x) = f_2(x)$.

14 Apply the rank-sum test to work Exercise 10. Average ranks for tie values occurring in A and B. Comment on the appropriateness of this test here as compared to the sign test.

15 Write down ten pairs of numbers such that it is clear that the rank-sum test would be very effective in rejecting the hypothesis that $f_1(x) = f_2(x)$, but for which the Student t test based on a normality assumption would not be effective.

16 Write down ten pairs of numbers such that it is clear that the t test based on a normality assumption would be very effective in rejecting the hypothesis that $f_1(x) = f_2(x)$, but for which the rank-sum test would not be effective.

17 Test the following set of measurements for randomness by means of the serial correlation test using a lag of 1. It is advisable to subtract a number such as 15 from each entry in order to simplify the computations.

11, 13, 12, 12, 14, 17, 16, 15, 15, 22, 18, 25, 26, 26, 27.

18 Test the following set of numbers for randomness by means of the serial correlation test using a lag of (a) 2, (b) 4.

1, 1, 3, 3, 1, 1, 3, 3, 1, 1, 3, 3, 1, 1, 3, 3.

19 Assuming that the serial correlation test based on R is equivalent to the same test based on the sample correlation coefficient r, show that the test is unaffected by subtracting the same constant from each observed value and by multiplying each observed value by the same constant.

20 Choose $f(x) = e^{-x}$, $x > 0$, and carry out the computations corresponding to those that were used to arrive at inequality (3). Compare your result with that given by inequality (2).

21 Choose 20 random digits from Table II and calculate a 90 percent confidence interval for σ^2,

(a) assuming normality and using χ^2;

(b) using (4). Compare your results and comment.

22 Suppose random samples of size 3 each from two distributions yielded the values 2, 7, 9 and 1, 3, 5, respectively. Determine the possible values of T, calculate the values of k_T, and then verify the critical values listed in Table IX.

Appendix—
Matrix Theory

The following approach to matrix theory is intended solely for the purpose of summarizing those parts of the theory that are needed in Chapters 4 and 5. It is presumed that the reader already has some familiarity with this material.

An n by k matrix A can be thought of as an array of numbers

$$
(1) \qquad A = \begin{bmatrix} a_{11} & \cdots & a_{1k} \\ \vdots & & \vdots \\ a_{n1} & \cdots & a_{nk} \end{bmatrix}
$$

having n rows and k columns. The number a_{ij} is called the i, j element of A. A matrix all of whose elements are zero is called a zero matrix. A is called a square matrix if $n = k$. A square matrix of the form

$$
I = \begin{bmatrix} 1 & 0 & \cdots & 0 \\ 0 & 1 & \cdots & 0 \\ \vdots & \vdots & & \vdots \\ 0 & 0 & \cdots & 1 \end{bmatrix}
$$

having 1's along the main diagonal and zeros elsewhere is called an identity matrix. The transpose A' of the matrix A is the k by n matrix defined by

$$
A' = \begin{bmatrix} a_{11} & \cdots & a_{n1} \\ \vdots & & \vdots \\ a_{1k} & \cdots & a_{nk} \end{bmatrix}.
$$

A square matrix A is called symmetric if $A' = A$, that is, if $a_{ij} = a_{ji}$ for all i, j.

If c is a real number, the matrix cA is defined as the n by k matrix whose i, j element is ca_{ij}, or

$$
cA = \begin{bmatrix} ca_{11} & \cdots & ca_{1k} \\ \vdots & & \vdots \\ ca_{n1} & \cdots & ca_{nk} \end{bmatrix}.
$$

Let B be an n by k matrix whose i, j element is b_{ij}. Then $A + B$ is defined as the n by k matrix whose i, j element is $a_{ij} + b_{ij}$, or

$$A + B = \begin{bmatrix} (a_{11} + b_{11}) & \cdots & (a_{1k} + b_{1k}) \\ \vdots & & \vdots \\ (a_{n1} + b_{n1}) & \cdots & (a_{nk} + b_{nk}) \end{bmatrix}.$$

Suppose now that B is a k by r matrix. Then the product AB of A and B is defined as the n by r matrix whose i, j element is

$$\sum_{m=1}^{k} a_{im} b_{mj}.$$

Thus,

$$AB = \begin{bmatrix} \sum_{m=1}^{k} a_{1m} b_{m1} & \cdots & \sum_{m=1}^{k} a_{1m} b_{mr} \\ \vdots & & \vdots \\ \sum_{m=1}^{k} a_{nm} b_{m1} & \cdots & \sum_{m=1}^{k} a_{nm} b_{mr} \end{bmatrix}.$$

Matrix multiplication is associative but not commutative. That is, $(AB)C = A(BC)$, but $AB \neq BA$ in general. The transpose of AB is given by

$$(AB)' = B'A'.$$

The matrix $A'A$ is symmetric, for

$$(A'A)' = A'A.$$

If A is an n by n matrix and I is the n by n identity matrix, then

$$AI = IA = A.$$

Let A and B both be $n \times n$ matrices. If $AB = I$, then $BA = I$. The matrix B that satisfies this equation is called the inverse of A and is denoted by A^{-1}. A square matrix that has an inverse is called nonsingular. If A and B are both nonsingular, then AB is nonsingular and

$$(AB)^{-1} = B^{-1}A^{-1}.$$

For

$$(B^{-1}A^{-1})(AB) = B^{-1}(A^{-1}A)B = B^{-1}B = I.$$

If A is nonsingular then A' is nonsingular and $(A')^{-1} = (A^{-1})'$. For

$$(A^{-1})'A' = (AA^{-1})' = I' = I.$$

In particular, if A is nonsingular and symmetric then $(A^{-1})' = (A')^{-1} = A^{-1}$, and hence A^{-1} is also symmetric.

For our purposes it is convenient to think of a vector having k components as a k by 1 matrix. If

$$\beta = \begin{bmatrix} \beta_1 \\ \vdots \\ \beta_k \end{bmatrix},$$

then β_j is called the jth component, or coordinate, of β, and is sometimes denoted by $(\beta)_j$. Specialization of the preceding matrix definitions immediately yields formulas for the sum of two vectors, a number times a vector, and a matrix times a vector. In particular, if A is given by (1), then $A\beta$ is defined by $(A\beta)_i = \sum_{j=1}^{k} a_{ij}\beta_j$, or

$$A\beta = \begin{bmatrix} \sum_{j=1}^{k} a_{1j}\beta_j \\ \vdots \\ \sum_{j=1}^{k} a_{nj}\beta_j \end{bmatrix}.$$

If I is the k by k identity matrix, then $I\beta = \beta$.

If α and β are two vectors having k components, then $\alpha'\beta$ is a 1 by 1 matrix. We can also think of $\alpha'\beta$ as the number

$$\alpha'\beta = \sum_{j=1}^{k} \alpha_j\beta_j.$$

In particular,

$$\beta'\beta = \sum_{j=1}^{k} \beta_j^2$$

is called the squared length of the vector β.

We can think of n-space R^n as a vector space consisting of all vectors having n components. Let $\mathcal{L}$ be a subset of R^n. We call $\mathcal{L}$ a subspace if whenever α and β are in $\mathcal{L}$ and c is any number, then $\alpha + \beta$ and $c\alpha$ are in $\mathcal{L}$.

Let $\alpha_1, \ldots, \alpha_k$ be k vectors in R^n. Let $\mathcal{L}$ be defined as the set of all linear combinations of the α's, that is, as all vectors of the form

$$c_1\alpha_1 + \cdots + c_k\alpha_k,$$

where $c_1, \ldots, c_k$ are arbitrary numbers. Then $\mathcal{L}$ is a subspace and we say that $\mathcal{L}$ is spanned by the vectors $\alpha_1, \ldots, \alpha_k$.

Let $\alpha_1, \ldots, \alpha_k$ be in R^n. These vectors are called linearly independent if no nontrivial linear combination of them equals the zero vector, that is,

$$c_1\alpha_1 + \cdots + c_k\alpha_k = 0,$$

where 0 denotes the zero vector, holds only when $c_1 = \cdots = c_k = 0$.

If a subspace $\mathcal{L}$ is spanned by k linearly independent vectors, $\alpha_1, \ldots, \alpha_k$, these vectors are called a basis of $\mathcal{L}$. Any vector μ in $\mathcal{L}$ can be expressed as a linear combination of the vectors in this basis

$$\mu = c_1\alpha_1 + \cdots + c_k\alpha_k,$$

where the numbers $c_1, \ldots, c_k$ are called the components, or coordinates, of μ with respect to this basis. For a given μ and a given basis, these components are uniquely determined.

If $\mathscr{L}$ is a subspace, one can choose a basis for $\mathscr{L}$ in a variety of ways, but all such bases must have the same number k of vectors. This common number k is called the dimension of $\mathscr{L}$. If $\mathscr{L}$ is k dimensional, vectors $\alpha_1, \ldots, \alpha_k$ in $\mathscr{L}$ span $\mathscr{L}$ if, and only if, they are linearly independent. If $\mathscr{L}$ is spanned by n vectors $\alpha_1, \ldots, \alpha_n$, $n \geq k$, then some subcollection of these vectors forms a basis for $\mathscr{L}$.

The space R^n is n dimensional and has as one basis the vectors $\delta_1, \ldots, \delta_n$, where

$$\delta_j = \begin{bmatrix} 0 \\ \vdots \\ 1 \\ \vdots \\ 0 \end{bmatrix}.$$

Here all components are 0 except for the jth component which is 1.

Let A be as in (1) and let $\alpha_1, \ldots, \alpha_k$ be vectors corresponding to the k columns of A, so that

$$\alpha_j = \begin{bmatrix} a_{1j} \\ \vdots \\ a_{nj} \end{bmatrix}, \qquad j = 1, \ldots, k.$$

If

$$\beta = \begin{bmatrix} \beta_1 \\ \vdots \\ \beta_k \end{bmatrix},$$

then

(2)
$$A\beta = \beta_1\alpha_1 + \cdots + \beta_k\alpha_k.$$

For

$$(A\beta)_i = \sum_{j=1}^{k} a_{ij}\beta_j = \sum_{j=1}^{k} \beta_j(\alpha_j)_i$$

$$= \left(\sum_{j=1}^{k} \beta_j\alpha_j \right)_i.$$

The rank of the matrix A is defined as the dimension of the subspace of R^n spanned by its column vectors $\alpha_1, \ldots, \alpha_k$. As β ranges over R^k, it follows from (2) that the collection $A\beta$ ranges over the subspace of R^n spanned by the column vectors of A and hence defines a subspace whose dimension equals the rank of A. The rank of the transpose A' of A equals the rank of A. An n by n matrix has rank n if and only if it is non-singular.

Vectors $\alpha_1, \ldots, \alpha_k$ are said to be orthonormal if

(3)
$$\alpha_i'\alpha_j = \begin{cases} 1, & i = j, \\ 0, & i \neq j. \end{cases}$$

Orthonormal vectors are necessarily linearly independent. For if (3) holds and

$$c_1\alpha_1 + \cdots + c_k\alpha_k = 0,$$

then for $j = 1, \ldots, k$,

$$0 = (c_1\alpha_1 + \cdots + c_k\alpha_k)'\alpha_j = c_1\alpha_1'\alpha_j + \cdots + c_k\alpha_k'\alpha_j$$

$$= c_j,$$

so that $c_1 = \cdots = c_k = 0$.

Let $\alpha_1, \ldots, \alpha_k$ be orthonormal vectors and let

$$\mu = c_1\alpha_1 + \cdots + c_k\alpha_k$$

for some numbers $c_1, \ldots, c_k$. Then the squared length of μ is given by

$$\mu'\mu = c_1^2 + \cdots + c_k^2.$$

For by (3)

$$\mu'\mu = (c_1\alpha_1 + \cdots + c_k\alpha_k)'(c_1\alpha_1 + \cdots + c_k\alpha_k)$$

$$= \sum_{i=1}^{k} \sum_{j=1}^{k} c_i c_j \alpha_i'\alpha_j = \sum_{i=1}^{k} c_i^2.$$

If μ is a linear combination of only $\alpha_1, \ldots, \alpha_{k-1}$, then $\mu'\alpha_k = 0$, because by (3)

$$\mu'\alpha_k = (c_1\alpha_1 + \cdots + c_{k-1}\alpha_{k-1})'\alpha_k = 0.$$

A basis consisting of orthonormal vectors is called an orthonormal basis. Every subspace $\mathscr{L}$ has an orthonormal basis. Moreover, if $\alpha_1, \ldots, \alpha_j$ are orthonormal vectors in $\mathscr{L}$, and j is less than the rank k of $\mathscr{L}$, then there exist $k - j$ additional vectors $\alpha_{j+1}, \ldots, a_k$ such that $\alpha_1, \ldots, \alpha_k$ is an orthonormal basis for $\mathscr{L}$

An n by n matrix A is called orthogonal if its columns

$$\alpha_j = \begin{bmatrix} a_{ij} \\ \vdots \\ a_{nj} \end{bmatrix}$$

form a set of orthonormal vectors. If A is orthogonal, then the calculation of $\alpha_i'\alpha_j$ shows that

$$\sum_{m=1}^{n} a_{mi}a_{mj} = \begin{cases} 1, & \text{if } i = j \\ 0, & \text{if } i \neq j. \end{cases}$$

It then follows that $A'A = I$. This implies that $A' = A^{-1}$, and hence also that $AA' = I$. If A is an orthogonal matrix, $A\beta$ has the same squared length as β for all β in R^n. For

$$(A\beta)'(A\beta) = \beta'A'A\beta = \beta'I\beta = \beta'\beta.$$

The reader is presumed to be familiar with the definition of the determinant, det A, of a square matrix A. Some of the important properties of determinants are that det $I = 1$, det $A' = $ det A, and det $AB = $ det $A \cdot$ det B. The square matrix A is nonsingular if and only if det $A \neq 0$. If A is nonsingular it follows from $AA^{-1} = I$ that

$$1 = \det I = \det A \cdot \det A^{-1},$$

and hence that det $A^{-1} = 1/\det A$. In particular, if A is orthogonal,

$$\det A = \det A' = \det A^{-1} = 1/\det A;$$

hence $(\det A)^2 = 1$ and det $A = 1$ or -1.

Let A be an n by k matrix of rank r. Then it is possible to delete $n - r$ rows and $k - r$ columns of A such that the resulting r by r submatrix has a nonzero determinant. It is impossible to find a larger square submatrix of A having a nonzero determinant. Let x be a vector with k components and c a vector with n components, then the vector equation $Ax = c$ represents n linear equations in k unknowns, namely,

$$a_{11}x_1 + \cdots + a_{1k}x_k = c_1$$
$$\vdots \qquad\qquad \vdots \qquad \vdots$$
$$a_{n1}x_1 + \cdots + a_{nk}x_k = c_n.$$

If A is of rank r, it is possible to find r of these equations that can be solved for r of the x's in terms of the c's and the remaining x's. If this solution satisfies the remaining $n - r$ equations, the equations are said to be consistent, otherwise inconsistent. As a special case, if $n = k$ and A is of rank n, there exists a unique solution of these equations given by $x = A^{-1}c$, because A then possesses an inverse and multiplying $Ax = c$ by A^{-1} gives $A^{-1}Ax = A^{-1}c$.

ILLUSTRATIONS

Let

$$A = \begin{bmatrix} 2 & 0 & 1 \\ 3 & 1 & -1 \\ 1 & 2 & 2 \end{bmatrix}, \quad B = \begin{bmatrix} 0 & 2 \\ 1 & 0 \\ -1 & 3 \end{bmatrix}, \quad \beta_1 = \begin{bmatrix} 1 \\ 0 \\ 1 \end{bmatrix}, \quad \beta_2 = \begin{bmatrix} 2 \\ 3 \\ 2 \end{bmatrix}.$$

Then

$$AB = \begin{bmatrix} -1 & 7 \\ 2 & 3 \\ 0 & 8 \end{bmatrix}, \quad (AB)' = \begin{bmatrix} -1 & 2 & 0 \\ 7 & 3 & 8 \end{bmatrix},$$

and

$$B'A' = \begin{bmatrix} 0 & 1 & -1 \\ 2 & 0 & 3 \end{bmatrix} \begin{bmatrix} 2 & 3 & 1 \\ 0 & 1 & 2 \\ 1 & -1 & 2 \end{bmatrix} = \begin{bmatrix} -1 & 2 & 0 \\ 7 & 3 & 8 \end{bmatrix} = (AB)'.$$

$$B'B = \begin{bmatrix} 2 & -3 \\ -3 & 13 \end{bmatrix},$$

which is symmetric.

$$A^{-1} = \begin{bmatrix} 4/13 & 2/13 & -1/13 \\ -7/13 & 3/13 & 5/13 \\ 5/13 & -4/13 & 2/13 \end{bmatrix},$$

and

$$AA^{-1} = \begin{bmatrix} 1 & 0 & 0 \\ 0 & 1 & 0 \\ 0 & 0 & 1 \end{bmatrix} = I.$$

$$\beta_1'\beta_2 = (1, 0, 1) \begin{bmatrix} 2 \\ 3 \\ 2 \end{bmatrix} = 4, \qquad A\beta_1 = \begin{bmatrix} 3 \\ 2 \\ 3 \end{bmatrix}.$$

If

$$c_1\beta_1 + c_2\beta_2 = c_1 \begin{bmatrix} 1 \\ 0 \\ 1 \end{bmatrix} + c_2 \begin{bmatrix} 2 \\ 3 \\ 2 \end{bmatrix} = \begin{bmatrix} c_1 \\ 0 \\ c_1 \end{bmatrix} + \begin{bmatrix} 2c_2 \\ 3c_2 \\ 2c_2 \end{bmatrix}$$

$$= \begin{bmatrix} c_1 + 2c_2 \\ 0 + 3c_2 \\ c_1 + 2c_2 \end{bmatrix} = \begin{bmatrix} 0 \\ 0 \\ 0 \end{bmatrix},$$

then

$$c_1 + 2c_2 = 0$$
$$0 + 3c_2 = 0$$
$$c_1 + 2c_2 = 0,$$

which implies that $c_2 = 0$ and $c_1 = 0$; hence β_1 and β_2 are linearly independent. They form a basis for $\mathscr{L}$, which is the plane in three dimensions passing through the origin and determined by these two vectors. The dimension of $\mathscr{L}$ is 2.

The vectors

$$\delta_1 = \begin{bmatrix} 1 \\ 0 \\ 0 \end{bmatrix}, \qquad \delta_2 = \begin{bmatrix} 0 \\ 1 \\ 0 \end{bmatrix}, \qquad \delta_2 = \begin{bmatrix} 0 \\ 0 \\ 1 \end{bmatrix},$$

which are the unit vectors along the x, y, and z axes, respectively, form an orthonormal basis for three-dimensional space because

$$\delta_i'\delta_j = \begin{cases} 1, & i = j \\ 0, & i \neq j, \end{cases}$$

and they are linearly independent. The column vectors of A, namely

$$\alpha_1 = \begin{bmatrix} 2 \\ 3 \\ 1 \end{bmatrix}, \qquad \alpha_2 = \begin{bmatrix} 0 \\ 1 \\ 2 \end{bmatrix}, \qquad \alpha_3 = \begin{bmatrix} 1 \\ -1 \\ 2 \end{bmatrix},$$

also span three-dimensional space because they are linearly independent.

For if

$$c_1\alpha_1 + c_2\alpha_2 + c_3\alpha_3 = \begin{bmatrix} 2c_1 + 0c_2 + c_3 \\ 3c_1 + c_2 - c_3 \\ c_1 + 2c_2 + 2c_3 \end{bmatrix} = \begin{bmatrix} 0 \\ 0 \\ 0 \end{bmatrix},$$

then

$$\begin{aligned} 2c_1 + 0c_2 + c_3 &= 0 \\ 3c_1 + c_2 - c_3 &= 0 \\ c_1 + 2c_2 + 2c_3 &= 0, \end{aligned}$$

which implies that $c_1 = c_2 = c_3 = 0$. Thus, the rank of A is 3, and A is a nonsingular matrix.

The matrix

$$P = \begin{bmatrix} \dfrac{1}{\sqrt{6}} & \dfrac{1}{\sqrt{3}} & \dfrac{1}{\sqrt{2}} \\ -\dfrac{1}{\sqrt{6}} & -\dfrac{1}{\sqrt{3}} & \dfrac{1}{\sqrt{2}} \\ \dfrac{2}{\sqrt{6}} & -\dfrac{1}{\sqrt{3}} & 0 \end{bmatrix}$$

is an orthogonal matrix because its column vectors satisfy

$$\alpha_i'\alpha_j = \begin{cases} 1, & i = j \\ 0, & i \neq j. \end{cases}$$

Then

$$P^{-1} = \begin{bmatrix} \dfrac{1}{\sqrt{6}} & -\dfrac{1}{\sqrt{6}} & \dfrac{2}{\sqrt{6}} \\ \dfrac{1}{\sqrt{3}} & -\dfrac{1}{\sqrt{3}} & -\dfrac{1}{\sqrt{3}} \\ \dfrac{1}{\sqrt{2}} & \dfrac{1}{\sqrt{2}} & 0 \end{bmatrix} = P',$$

and det $P = 1$.

Let

$$C = \begin{bmatrix} 2 & 0 & -2 & 2 \\ 3 & 1 & -1 & 2 \\ 1 & 1 & 1 & 0 \end{bmatrix}.$$

Then C is of rank 2, because if $\alpha_1, \alpha_2, \alpha_3,$ and α_4 denote its column vectors, then $\alpha_3 = -\alpha_1 + 2\alpha_2$ and $\alpha_4 = \alpha_1 - \alpha_2$, and therefore the two linearly independent column vectors α_1 and α_2 span the same space as that spanned by all four column vectors. The 2 by 2 matrix obtained by deleting the last row and the last two columns of C is nonsingular with determinant value 2. The three homogeneous equations of the form $Cx = 0$, or

$$2x_1 + 0x_2 - 2x_3 + 2x_4 = 0$$
$$3x_1 + x_2 - x_3 + 2x_4 = 0$$
$$x_1 + x_2 + x_3 + 0x_4 = 0,$$

can therefore be solved for x_1 and x_2 in terms of x_3 and x_4 by solving the first two equations. Thus,

$$\begin{cases} 2x_1 + 0x_2 = 2x_3 - 2x_4 \\ 3x_1 + x_2 = x_3 - 2x_4 \end{cases} \quad \text{have the solution} \quad \begin{cases} x_1 = x_3 - x_4 \\ x_2 = -2x_3 + x_4. \end{cases}$$

These values satisfy the third equation: therefore the equations are consistent.

Answers

CHAPTER 1

1. $c^2 + \theta^2(c-1)^2$. **2.** $d(x) = x$.

3. (a) $\theta(1-\theta)/2$, (b) $(2\theta^2 - 2\theta + 1)/16$, (c) $(x+1)/4$, (d) No.

4. (a) $(4c^2 - 2c + 1)/3$, (b) $c = 1/4$.

5. (a) $\theta(1-\theta)/n$, (b) $1/6n$. **6.** (a) $1/n$, (b) $1/n$.

7. (a) θ, (b) α/λ.

8. (a) $\mathscr{R}(1/4, d_1) = 0$, $\mathscr{R}(3/4, d_1) = 1/4$, $\mathscr{R}(1/4, d_2) = 1/4$, $\mathscr{R}(3/4, d_2) = 0$,
$\mathscr{R}(1/4, d_3) = 9/64$, $\mathscr{R}(3/4, d_3) = 15/64$; (b) d_3.

9. $\mathscr{R}(0, d_1) = 0$, $\mathscr{R}(1, d_1) = 1/2$, $\mathscr{R}(0, d_2) = 1/2$, $\mathscr{R}(1, d_2) = 3/4$; hence d_1 is minimax.

CHAPTER 2

2. $c = 2$.

3. (a) $c = 1$, (b) no solution to $c\theta(\theta + 1) = \theta$ for all $\theta > 0$.

4. $\displaystyle\sum_{i=1}^{n} a_i = 1$.

5. $b(p) = (1/2 - p)/(\sqrt{n} + 1)$.

7. N trials will be required if $x - 1$ successes have occurred in $N - 1$ trials and the next trial produces a success.

9. $a_i = 1/n$, $i = 1, \ldots, n$. **10.** $e(Z) = 2/n$.

12. $2/n$.

13. Upper limit is a function of θ; therefore equality obtained by differentiating integral is incorrect.

14. $E(Z) = \theta$, $V(Z) = 2\theta^2/n$. **17.** $1/\bar{X}$.

18. $EX = 1/\theta = \varphi$ and $\hat{\varphi} = \bar{X}$; hence reciprocal estimates for reciprocal parameters.

19. $(\bar{X} + 1)^{-1}$. **21.** $-(n + \sum \log X_i)/\sum \log X_i$.

22. $\sqrt{\sum X_i^2/n}$. **23.** $\sum X_i^2/n$.

24. $\bar{x} = 9.86$, $s^2 = 10.8$.   **29.** $k = n/(n + 2)$.

30. $2\bar{x}$.   **32.** (a) $2\bar{x}^2/\pi$,   (b) $\sum x_i^2/2n$.

33. $p = \dfrac{\bar{x} - s^2}{\bar{x}}$, $n = \dfrac{\bar{x}^2}{\bar{x} - s^2}$.

36. Moments do not exist.

38. $\dfrac{1}{n} \sum\limits_{i=1}^{n} \dfrac{1}{\sqrt{x_i}}$, $\sqrt{\dfrac{\pi}{\bar{x}}}$

39. $(18.69, 21.31)$.   **40.** 100.

42. $(12.7, 36.9)$.   **45.** $(2.35, 10.9)$.   **46.** 1537.

47. $(7.15, 8.85)$.   **48.** $(.17, .43)$.   **51.** $(3.1, 5.6)$.

52. $(S + 2)/(n + 4)$, where $S =$ total number of successes.

53. (a) e^{-x},   (b) $e^{x-\theta}$, $\theta \geq x$,   (c) $x + 1$.

54. $(S + 1)/(n + 1)$ where $S = \sum X_i$.

55. $\dfrac{2}{x} - \dfrac{x}{e^x - x - 1}$.

56. (a) $\mathscr{R}(\theta, d_1) = \theta^2$, $\mathscr{R}(\theta, d_2) = \theta^2 + 1$; hence d_1 is minimax.
   (b) $r(\pi, d_1) = 5/2$, $r(\pi, d_2) = 7/2$; hence d_1 is Bayes solution.

57. (a) $\mathscr{R}(\theta, d_1) = \theta/2$, $r(\pi, d_1) = 1/4$,   (b) $\mathscr{R}(\theta, d_2) = \theta - \theta^2/3$, $r(\pi, d_2) = 7/18$.

58. $(\alpha + a)/(x + \beta)$.

59. (a) $2/3x$,   (b) same as in (a) because (a) is linear.

60. (a) $(x + 4/3)/(2x + 2)$,   (b) $d(x) = -(x/13) + 17/26$,   (c) excellent approximation.

61. $f(p \mid s) = c(m, s)p^s(1 - p)^{m-s}$; $E(p \mid S) = (s + S + 1)/(m + n + 2)$; the same.

CHAPTER 3

1. $\alpha = 3/4$, $\beta = 4/9$.

2. (a) $\alpha = 1/2$, $\beta = 1/4$,   (b) $\alpha = 0$, $\beta = 3/4$.

3. $\alpha = 1/56$, $\beta = 46/56$.

4. (a) $x = 2, 3, 4$, $x = 2, 3, 5$, $x = 4, 5$, $x = 1, 2, 6$, $x = 3, 6$;   (b) $x = 2, 3, 5$.

5. $x = 0$, $\beta = 26/27$; $x = 3$, $\beta = 19/27$; hence choose $x = 3$ with $\alpha = 1/8$.

6. $f(x \mid H_0) = 1/9$, $f(x \mid H_1) = (10 - x)/45$; $\alpha = 5/9$, $\beta = 2/9$.

7. (a) $x = 0$,   (b) $x = 1$,   (c) $x = 2$,   (d) $x = 0, 2$,   (e) gives minimum
   $\alpha + \beta = 1/4$.

8. $\sum \log x_i \leq a$.   **9.** $\sum x_i \geq a$.

10. If $a(\theta_1) > a(\theta_0)$, then $\sum b(x_i) \geq a$; if $a(\theta_1) < a(\theta_0)$, then $\sum b(x_i) \leq a$.

11. $\alpha = 1/4$, critical region is $x = 1$; $\alpha = 7/16$, critical region is $x = 1, 2$.

12. $\tau = 2.8$; hence reject H_0.

13. $\chi^2 = 44.4$, $\chi_0^2 = 31.4$; hence reject H_0.

14. $\chi^2 = 15.6$, $\chi_0^2 = 16.79$; hence reject H_0.

16. $\tau = 2.24$, $\tau_0 = 1.28$; hence reject H_0.

17. $\tau = 1.6$, $\tau_0 = 1.28$, hence reject H_0.

18. $\tau = -1.67$, $\tau_0 = -1.96$; hence accept H_0.

19. $\tau = -1.2$; hence reject claim.   **21.** $\beta = P(\tau < -3.72) = .0004$.

22. Critical region $X \geq 24.04$, $\beta = .50$.

23. .79.   **24.** 9.   **25.** 63.

26. $c = 0 : \alpha = .50$, $\beta = .025$; $c = 1/2 : \alpha = .16$, $\beta = .16$; $c = 1 : \alpha = .025$, $\beta = .50$; $c = \infty : \alpha = 0$, $\beta = 1$; $c = -\infty : \alpha = 1$, $\beta = 0$.

27. $P(0) = .05$, $P(\pm 1/2) = .17$, $P(\pm 1) = .52$, $P(\pm 2) = .98$.

28. $P(p) = \binom{10 - 10p}{4} / \binom{10}{4} + \binom{10p}{1}\binom{10 - 10p}{3} / \binom{10}{4}$.

29. (a) $P(p) = p^3$,   (b) $P(p) = p^3 + 3p^2(1 - p)$.

30. $P(1/2) = .29$, $P(3/4) = P(1/4) = .68$, $P(1) = P(0) = 1$.

31. Yes.   **32.** Yes.

33. Choose as critical region $\bar{X} \geq 1.64\sigma_{\bar{X}}$; then $\alpha = .05$ and $\beta(1) = .36$. $\beta(1) = .48$ for Exercise 27; hence not U.M.P.

34. (a) $x \geq .9$,   (b) Yes.

37. $(\log 2/9 + n \log 1.2)7200/11 < \sum x_i^2 < (\log 8 + n \log 1.2)7200/11$.

38. $\log 1/9 + n \log 4/3 < \sum x_i < \log 9 + n \log 4/3$.

39. $\left[\log \dfrac{\beta}{1 - \alpha} + n(\mu_1 - \mu_0)\right] / \log \dfrac{\mu_1}{\mu_0} < \sum x_i < \left[\log \dfrac{1 - \beta}{\alpha} + n(\mu_1 - \mu_0)\right] / \log \dfrac{\mu_1}{\mu_0}$

40. $-32(\log 9 + n \log 2) < \sum (x_i - 12)^2 - 4\sum (x_i - 10)^2 < 32(\log 9 - n \log 2)$.

41. 14.   **42.** 70.   **43.** 30.

45. $t = -1.55$; hence accept H_0.

47. (16.2, 23.8).   **49.** $t = .81$; hence accept H_0.

50. (a) $t = 3.1$; hence reject H_0;   (b) $.5 < \mu_A - \mu_B < 2.5$.

51. $-2.2 < \mu_x - \mu_y < 6.2$.   **53.** $F = 2.25$; hence accept H_0.

54. $F = 2.13$; hence accept H_0.   **56.** Critical region $(\bar{x} - \mu_0)^2 > c$.

57. Critical region $\sum (x_i - \mu)^2 < c_1$ and $\sum (x_i - \mu)^2 > c_2 > c_1$.

58. $\chi^2 = 11.7$; hence reject theory.   **61.** $\chi^2 = 3.3$; hence accept H_0.

62. $\chi^2 = 2.5$; hence accept H_0.   **63.** $\chi^2 =$ same; hence accept H_0.

64. $\chi^2 = .32$; hence accept H_0.   **65.** $\chi^2 = 1.0$; hence accept H_0.

66. $\chi^2 = 16$; hence reject independence.

68. (a) $A_1 : x \leq -1/2$, $A_2 : -1/2 < x < 1/2$,   (b) $A_1 : x \leq -1/2 - \log 2$, $A_2 : -1/2 - \log 2 < x < 1/2 - \log 2$, $A_3 \geq 1/2 - \log 2$.

69. $A_1 : x \geq \log 2$, $A_2 : \log 3/2 < x < \log 2$, $A_3 : x \leq \log 3/2$.

70. $A_1 : y \leq -x$, $x \leq -1/2 \log 2$, $A_2 : y > -x$, $y \geq x + \log 2$, $A_3 : x > -1/2 \log 2$, $y < x + \log 2$.

CHAPTER 4

1. $y = 3.2 + 1/2(x - 3)$, or $y = 1.7 + 1/2x$.

2. $y = -274.4 + 6.288x$.   **3.** (a) $y = 1/3 + 4/3x$, (b) $y = 7/5x$.

4. $a \sum x_i^2 + bn = \sum x_i y_i$, $an + b \sum 1/x_i^2 = \sum y_i/x_i$.

5. No. $z = \alpha + \log x + \beta x^2$ where $z = \log y$, $\alpha = \log a$, $\beta = -b$ and least squares estimates of α and β.

6. $a = 1.42$, $b = 1.00$.

7. $y = 3.249 - .172(x - .4) - .754(x - .4)^2$.

8. $y = 1/2x$.   **9.** $E(Y \mid x) = 1/x$.   **10.** $E(Y \mid x) = \mu_y$.

11. $y = 3.380x_1 + .00358x_2$, or $Y = 3.380x_1 + .00358x_2 + 9.326$.

12. $y = .39x_1 + .25x_2$.

17. $V(a) = \sigma^2/5$, $V(b) = \sigma^2/10$ if $y = a + b(x - 3)$.

18. $V(a) = 32.54\sigma^2$, $V(b) = .0069\sigma^2$.

19. (a) $V(a) = \sigma^2/18$, $V(b) = \sigma^2/180$; (b) $V(b) = \sigma^2/300$.

20. (a) $y = 13/6x_1 + 5/12x_2 + 407/18$.
    (b) $V(\hat{\beta}_1) = \sigma^2/6$, $V(\hat{\beta}_2) = \sigma^2/24$, $V(\hat{\beta}_3) = 17\sigma^2/18$.
    (c) $\hat{\beta}_2$ most, and $\hat{\beta}_3$ least accurate. Both beneficial.

21. $y = -1.416 + .140x_1 - .005x_2$.

22. $V(a_1) = .005\sigma^2$, $V(a_2) = .032\sigma^2$.

23. $V(\hat{\beta}_1) = .94\sigma^2$, $V(\hat{\beta}_2) = .004\sigma^2$, $V(\hat{\beta}_3) = .001\sigma^2$.

24. $s^2 = 2.66$; $13/6 \pm 1.3$, $5/12 \pm .67$, $407/18 \pm 3.2$.

26. $V(z) = \sigma^2 \left[\dfrac{1}{n} + \dfrac{(x - \bar{x})^2}{\sum (x_i - \bar{x})^2} \right]$.

27. Choose half x_i at -1 and half at 1 to maximize $\sum (x_i - x)^2$, assuming n even.

28. $P_0(x) = 1/2$, $P_1(x) = (x - 3/2)/\sqrt{5}$, $P_2(x) = (x^2 - 3x + 1)/2$.

29. $a_r = \displaystyle\sum_{i=1}^{n} y_i P_r(x_i)$.

31. $\bar{y}_1. - \bar{y} = -1.92$, $\bar{y}_2. - \bar{y} = 2.83$, $\bar{y}_3. - \bar{y} = -.92$, $\bar{y}_{.1} - \bar{y} = -3.33$,
    $\bar{y}_{.2} - \bar{y} = 2.33$, $\bar{y}_{.3} - \bar{y} = -1.33$, $\bar{y}_{.4} - \bar{y} = 2.33$.

32. $\bar{y}_1. - \bar{y} = .25$, $\bar{y}_2. - \bar{y} = 4.25$, $\bar{y}_3. - \bar{y} = -3.50$, $\bar{y}_4. - \bar{y} = -1.75$,
    $\bar{y}_5. - \bar{y} = .75$, $\bar{y}_{.1} - \bar{y} = -2.15$, $\bar{y}_{.2} - \bar{y} = -1.75$, $\bar{y}_{.3} - \bar{y} = 7.45$,
    $\bar{y}_{.4} - \bar{y} = -3.55$.

33. 2.5.   **34.** 12.

36. $V(\bar{Y}_{.j} - \bar{Y}) = (c - 1)\sigma^2/rc$.

37. $-4.6 < b_1 < -2.0$.   **38.** $E(\bar{Y}_{.j}) = \mu$, $V(\bar{Y}_{.j}) = \sigma^2/r$.

41. $V = \sigma^2(r - 1)(c - 1)/rc$.

CHAPTER 5

1. Ranks are 2 and 1; $n \geq 2$ and at least two unequal x's.

2. Ranks are 2 and 1.   **4.** Rank is 3.

7. $\mathscr{L}$ is the y_1, y_2 plane; $\mathscr{L}_0$ is the y_1 axis.

10. $1/2\pi \exp \left[-1/2[(z_1 - 3/2\sqrt{2})^2 + (z_2 - 1/2\sqrt{2})^2]\right]$.

12. $F = 58$, $F_0 = 4.2$; hence reject H_0.

13. $F = 2.15$; hence accept H_0.

14. $F = .9$, $F_0 = 4.2$; hence accept H_0.

15. $F = .88$; hence accept H_0.

16. $F = 1.6$, $F_0 = 5.99$; hence accept H_0.

17. $F = 6$, $F_0 = 5.14$; hence reject H_0.

18. $F \geq 200$; hence reject H_0.

20. $.56 \pm 1.8\sqrt{1170/4550(11)}$; hence (.28, .84).

21. $1/9 \pm 3.18\sqrt{.34/216(3)}$; hence (.038, .184).

22. $13/6 \pm 1.3$ and $5/12 \pm .67$; hence correct intervals are much wider.

25. $F = 6.6$; hence reject H_0.

26. $F = 10.0$; hence reject H_0; $F = 9.5$; hence reject H_0.

27. $F = 2.8$; hence accept H_0; $F = 10.5$; hence reject H_0.

30. $F = 7.2$; hence reject H_0. Earlier $F = 10.5$. Only Exercise 27 model appropriate because rows do not represent random sampling for column values.

CHAPTER 6

2. Band given by $S_{20}(x) \pm .23$. $F(x)$ lies well inside the band.

4. Maximum difference occurs at $x = .25$. Its value there is .151. Since $D_{20}^{.05} = .29$, H_0 is accepted.

5. Maximum difference occurs at $x = 34.7$. Its value there is .442. Since $D_{10}^{.20} = .32$, H_0 is rejected.

7. $\tau = 2.53$; hence reject $H_0: p = 1/2$.

8. $\tau = 1.52$; hence accept $H_0: p = 1/2$.

9. $\tau = -2.06$; hence reject H_0. Test based on normality and $\sigma = 1$ gives $\tau = -1.72$; hence reject H_0.

10. $\tau = 2.14$; hence reject $H_0: \xi_1 = \xi_2$.

12. $T = 197$, $\tau = 3.7$; hence reject no shift hypothesis.

14. $T = 191$, $\tau = -.6$; hence accept $H_0: f_1(x) = f_2(x)$. A poor test since it ignores the matching.

17. Subtract 20. $R = -327$, $\tau = -3.1$; hence reject randomness.

18. (a) $\tau = -3.7$; hence reject randomness, (b) $\tau = -4.3$; hence reject randomness.

20. The acceptance interval will be approximately twice as long.

Tables

Table I Squares and square roots

Number	Square	Square Root	Number	Square	Square Root
1	1	1.000	41	16 81	6.403
2	4	1.414	42	17 64	6.481
3	9	1.732	43	18 49	6.557
4	16	2.000	44	19 36	6.633
5	25	2.236	45	20 25	6.708
6	36	2.449	46	21 16	6.782
7	49	2.646	47	22 09	6.856
8	64	2.828	48	23 04	6.928
9	81	3.000	49	24 01	7.000
10	1 00	3.162	50	25 00	7.071
11	1 21	3.317	51	26 01	7.141
12	1 44	3.464	52	27 04	7.211
13	1 69	3.606	53	28 09	7.280
14	1 96	3.742	54	29 16	7.348
15	2 25	3.873	55	30 25	7.416
16	2 56	4.000	56	31 36	7.483
17	2 89	4.123	57	32 49	7.550
18	3 24	4.243	58	33 64	7.616
19	3 61	4.359	59	34 81	7.681
20	4 00	4.472	60	36 00	7.746
21	4 41	4.583	61	37 21	7.810
22	4 84	4.690	62	38 44	7.874
23	5 29	4.796	63	39 69	7.937
24	5 76	4.899	64	40 96	8.000
25	6 25	5.000	65	42 25	8.062
26	6 76	5.099	66	43 56	8.124
27	7 29	5.196	67	44 89	8.185
28	7 84	5.292	68	46 24	8.246
29	8 41	5.385	69	47 61	8.307
30	9 00	5.477	70	49 00	8.367
31	9 61	5.568	71	50 41	8.426
32	10 24	5.657	72	51 84	8.485
33	10 89	5.745	73	53 29	8.544
34	11 56	5.831	74	54 76	8.602
35	12 25	5.916	75	56 25	8.660
36	12 96	6.000	76	57 76	8.718
37	13 69	6.083	77	59 29	8.775
38	14 44	6.164	78	60 84	8.832
39	15 21	6.245	79	62 41	8.888
40	16 00	6.325	80	64 00	8.944

Table I   Squares and square roots

Number	Square	Square Root	Number	Square	Square Root
81	65 61	9.000	121	1 46 41	11.000
82	67 24	9.055	122	1 48 84	11.045
83	68 89	9.110	123	1 51 29	11.091
84	70 56	9.165	124	1 53 76	11.136
85	72 25	9.220	125	1 56 25	11.180
86	73 96	9.274	126	1 58 76	11.225
87	75 69	9.327	127	1 61 29	11.269
88	77 44	9.381	128	1 63 84	11.314
89	79 21	9.434	129	1 66 41	11.358
90	81 00	9.487	130	1 69 00	11.402
91	82 81	9.539	131	1 71 61	11.446
92	84 64	9.592	132	1 74 24	11.489
93	86 49	9.644	133	1 76 89	11.533
94	88 36	9.695	134	1 79 56	11.576
95	90 25	9.747	135	1 82 25	11.619
96	92 16	9.798	136	1 84 96	11.662
97	94 09	9.849	137	1 87 69	11.705
98	96 04	9.899	138	1 90 44	11.747
99	98 01	9.950	139	1 93 21	11.790
100	1 00 00	10.000	140	1 96 00	11.832
101	1 02 01	10.050	141	1 98 81	11.874
102	1 04 04	10.100	142	2 01 64	11.916
103	1 06 09	10.149	143	2 04 49	11.958
104	1 08 16	10.198	144	2 07 36	12.000
105	1 10 25	10.247	145	2 10 25	12.042
106	1 12 36	10.296	146	2 13 16	12.083
107	1 14 49	10.344	147	2 16 09	12.124
108	1 16 64	10.392	148	2 19 04	12.166
109	1 18 81	10.440	149	2 22 01	12.207
110	1 21 00	10.488	150	2 25 00	12.247
111	1 23 21	10.536	151	2 28 01	12.288
112	1 25 44	10.583	152	2 31 04	12.329
113	1 27 69	10.630	153	2 34 09	12.369
114	1 29 96	10.677	154	2 37 16	12.410
115	1 32 25	10.724	155	2 40 25	12.450
116	1 34 56	10.770	156	2 43 36	12.490
117	1 36 89	10.817	157	2 46 49	12.530
118	1 39 24	10.863	158	2 49 64	12.570
119	1 41 61	10.909	159	2 52 81	12.610
120	1 44 00	10.954	160	2 56 00	12.649

Table I Squares and square roots

Number	Square	Square Root	Number	Square	Square Root
161	2 59 21	12.689	201	4 04 01	14.177
162	2 62 44	12.728	202	4 08 04	14.213
163	2 65 69	12.767	203	4 12 09	14.248
164	2 68 96	12.806	204	4 16 16	14.283
165	2 72 25	12.845	205	4 20 25	14.318
166	2 75 56	12.884	206	4 24 36	14.353
167	2 78 89	12.923	207	4 28 49	14.387
168	2 82 24	12.961	208	4 32 64	14.422
169	2 85 61	13.000	209	4 36 81	14.457
170	2 89 00	13.038	210	4 41 00	14.491
171	2 92 41	13.077	211	4 45 21	14.526
172	2 95 84	13.115	212	4 49 44	14.560
173	2 99 29	13.153	213	4 53 69	14.595
174	3 02 76	13.191	214	4 57 96	14.629
175	3 06 25	13.229	215	4 62 25	14.663
176	3 09 76	13.266	216	4 66 56	14.697
177	3 13 29	13.304	217	4 70 89	14.731
178	3 16 84	13.342	218	4 75 24	14.765
179	3 20 41	13.379	219	4 79 61	14.799
180	3 24 00	13.416	220	4 84 00	14.832
181	3 27 61	13.454	221	4 88 41	14.866
182	3 31 24	13.491	222	4 92 84	14.900
183	3 34 89	13.528	223	4 97 29	14.933
184	3 38 56	13.565	224	5 01 76	14.967
185	3 42 25	13.601	225	5 06 25	15.000
186	3 45 96	13.638	226	5 10 76	15.033
187	3 49 69	13.675	227	5 15 29	15.067
188	3 53 44	13.711	228	5 19 84	15.100
189	3 57 21	13.748	229	5 24 41	15.133
190	3 61 00	13.784	230	5 29 00	15.166
191	3 64 81	13.820	231	5 33 61	15.199
192	3 68 64	13.856	232	5 38 24	15.232
193	3 72 49	13.892	233	5 42 89	15.264
194	3 76 36	13.928	234	5 47 56	15.297
195	3 80 25	13.964	235	5 52 25	15.330
196	3 84 16	14.000	236	5 56 96	15.362
197	3 88 09	14.036	237	5 61 69	15.395
198	3 92 04	14.071	238	5 66 44	15.427
199	3 96 01	14.107	239	5 71 21	15.460
200	4 00 00	14.142	240	5 76 00	15.492

Table I Squares and square roots

Number	Square	Square Root	Number	Square	Square Root
241	5 80 81	15.524	281	7 89 61	16.763
242	5 85 64	15.556	282	7 95 24	16.793
243	5 90 49	15.588	283	8 00 89	16.823
244	5 95 36	15.620	284	8 06 56	16.852
245	6 00 25	15.652	285	8 12 25	16.882
246	6 05 16	15.684	286	8 17 96	16.912
247	6 10 09	15.716	287	8 23 69	16.941
248	6 15 04	15.748	288	8 29 44	16.971
249	6 20 01	15.780	289	8 35 21	17.000
250	6 25 00	15.811	290	8 41 00	17.029
251	6 30 01	15.843	291	8 46 81	17.059
252	6 35 04	15.875	292	8 52 64	17.088
253	6 40 09	15.906	293	8 58 49	17.117
254	6 45 16	15.937	294	8 64 36	17.146
255	6 50 25	15.969	295	8 70 25	17.176
256	6 55 36	16.000	296	8 76 16	17.205
257	6 60 49	16.031	297	8 82 09	17.234
258	6 65 64	16.062	298	8 88 04	17.263
259	6 70 81	16.093	299	8 94 01	17.292
260	6 76 00	16.125	300	9 00 00	17.321
261	6 81 21	16.155	301	9 06 01	17.349
262	6 86 44	16.186	302	9 12 04	17.378
263	6 91 69	16.217	303	9 18 09	17.407
264	6 96 96	16.248	304	9 24 16	17.436
265	7 02 25	16.279	305	9 30 25	17.464
266	7 07 56	16.310	306	9 36 36	17.493
267	7 12 89	16.340	307	9 42 49	17.521
268	7 18 24	16.371	308	9 48 64	17.550
269	3 23 61	16.401	309	9 54 81	17.578
270	7 29 00	16.432	310	9 61 00	17.607
271	7 34 41	16.462	311	9 67 21	17.635
272	7 39 84	16.492	312	9 73 44	17.664
273	7 45 29	16.523	313	9 79 69	17.692
274	7 50 76	16.553	314	9 85 96	17.720
275	7 56 25	16.583	315	9 92 25	17.748
276	7 61 76	16.613	316	9 98 56	17.776
277	7 67 29	16.643	317	10 04 89	17.804
278	7 72 84	16.673	318	10 11 24	17.833
279	7 78 41	16.703	319	10 17 61	17.861
280	7 84 00	16.733	320	10 24 00	17.889

Table I   Squares and square roots

Number	Square	Square Root	Number	Square	Square Root
321	10 30 41	17.916	361	13 03 21	19.000
322	10 36 84	17.944	362	13 10 44	19.026
323	10 43 29	17.972	363	13 17 69	19.053
324	10 49 76	18.000	364	13 24 96	19.079
325	10 56 25	18.028	365	13 32 25	19.105
326	10 62 76	18.055	366	13 39 56	19.131
327	10 69 29	18.083	367	13 46 89	19.157
328	10 75 84	18.111	368	13 54 24	19.183
329	10 82 41	18.138	369	13 61 61	19.209
330	10 89 00	18.166	370	13 69 00	19.235
331	10 95 61	18.193	371	13 76 41	19.261
332	11 02 24	18.221	372	13 83 84	19.287
333	11 08 89	18.248	373	13 91 29	19.313
334	11 15 56	18.276	374	13 98 76	19.339
335	11 22 25	18.303	375	14 06 25	19.363
336	11 28 96	18.330	376	14 13 76	19.391
337	11 35 69	18.358	377	14 21 29	19.416
338	11 42 44	18.385	378	14 28 84	19.442
339	11 49 21	18.412	379	14 36 41	19.468
340	11 56 00	18.439	380	14 44 00	19.494
341	11 62 81	18.466	381	14 51 61	19.519
342	11 69 64	18.493	382	14 59 24	19.545
343	11 76 49	18.520	383	14 66 89	19.570
344	11 83 36	18.547	384	14 74 56	19.596
345	11 90 25	18.574	385	14 82 25	19.621
346	11 97 16	18.601	386	14 89 96	19.647
347	12 04 09	18.628	387	14 97 69	19.672
348	12 11 04	18.655	388	15 05 44	19.698
349	12 18 01	18.682	389	15 13 21	19.723
350	12 25 00	18.708	390	15 21 00	19.748
351	12 32 01	18.735	391	15 28 81	19.774
352	12 39 04	18.762	392	15 36 64	19.799
353	12 46 09	18.788	393	15 44 49	19.824
354	12 53 16	18.815	394	15 52 36	19.849
355	12 60 25	18.841	395	15 60 25	19.875
356	12 67 36	18.868	396	15 68 16	19.900
357	12 74 49	18.894	397	15 76 09	19.925
358	12 81 64	18.921	398	15 84 04	19.950
359	12 88 81	18.947	399	15 92 01	19.975
360	12 96 00	18.974	400	16 00 00	20.000

Table I   Squares and square roots

Number	Square	Square Root	Number	Square	Square Root
401	16 08 01	20.025	441	19 44 81	21.000
402	16 16 04	20.050	442	19 53 64	21.024
403	16 24 09	20.075	443	19 62 49	21.048
404	16 32 16	20.100	444	19 71 36	21.071
405	16 40 25	20.125	445	19 80 25	21.095
406	16 48 36	20.149	446	19 89 16	21.119
407	16 56 49	20.174	447	19 98 09	21.142
408	16 64 64	20.199	448	20 07 04	21.166
409	16 72 81	20.224	449	20 16 01	21.190
410	16 81 00	20.248	450	20 25 00	21.213
411	16 89 21	20.273	451	20 34 01	21.237
412	16 97 44	20.298	452	20 43 04	21.260
413	17 05 69	20.322	453	20 52 09	21.284
414	17 13 96	20.347	454	20 61 16	21.307
415	17 22 25	20.372	455	20 70 25	21.331
416	17 30 56	20.396	456	20 79 36	21.354
417	17 38 89	20.421	457	20 88 49	21.378
418	17 47 24	20.445	458	20 97 64	21.401
419	17 55 61	20.469	459	21 06 81	21.424
420	17 64 00	20.494	460	21 16 00	21.448
421	17 72 41	20.518	461	21 25 21	21.471
422	17 80 84	20.543	462	21 34 44	21.494
423	17 89 29	20.567	463	21 43 69	21.517
424	17 97 76	20.591	464	21 52 96	21.541
425	18 06 25	20.616	465	21 62 25	21.564
426	18 14 76	20.640	466	21 71 56	21.587
427	18 23 29	20.664	467	21 80 89	21.610
428	18 31 84	20.688	468	21 90 24	21.633
429	18 40 41	20.712	469	21 99 61	21.656
430	18 49 00	20.736	470	22 09 00	21.679
431	18 57 61	20.761	471	22 18 41	21.703
432	18 66 24	20.785	472	22 27 84	21.726
433	18 74 89	20.809	473	22 37 29	21.749
434	18 83 56	20.833	474	22 46 76	21.772
435	18 92 25	20.857	475	22 56 25	21.794
436	19 00 96	20.881	476	22 65 76	21.817
437	19 09 69	20.905	477	22 75 29	21.840
438	19 18 44	20.928	478	22 84 84	21.863
439	19 27 21	20.952	479	22 94 41	21.886
440	19 36 00	20.976	480	23 04 00	21.909

Table I   Squares and square roots

Number	Square	Square Root	Number	Square	Square Root
481	23 13 61	21.932	521	27 14 41	22.825
482	23 23 24	21.954	522	27 24 84	22.847
483	23 32 89	21.977	523	27 35 29	22.869
484	23 42 56	22.000	524	27 45 76	22.891
485	23 52 25	22.023	525	27 56 25	22.913
486	23 61 96	22.045	526	27 66 76	22.935
487	23 71 69	22.068	527	27 77 29	22.956
488	23 81 44	22.091	528	27 87 84	22.978
489	23 91 21	22.113	529	27 98 41	23.000
490	24 01 00	22.136	530	28 09 00	23.022
491	24 10 81	22.159	531	28 19 61	23.043
492	24 20 64	22.181	532	28 30 24	23.065
493	24 30 49	22.204	533	28 40 89	23.087
494	24 40 36	22.226	534	28 51 56	23.108
495	24 50 25	22.249	535	28 62 25	23.130
496	24 60 16	22.271	536	28 72 96	23.152
497	24 70 09	22.293	537	28 83 69	23.173
498	24 80 04	22.316	538	28 94 44	23.195
499	24 90 01	22.338	539	29 05 21	23.216
500	25 00 00	22.361	540	29 16 00	23.238
501	25 10 01	22.383	541	29 26 81	23.259
502	25 20 04	22.405	542	29 37 64	23.281
503	25 30 09	22.428	543	29 48 49	23.302
504	25 40 16	22.450	544	29 59 36	23.324
505	25 50 25	22.472	545	29 70 25	23.345
506	25 60 36	22.494	546	29 81 16	23.367
507	25 70 49	22.517	547	29 92 09	23.388
508	25 80 64	22.539	548	30 03 04	23.409
509	25 90 81	22.561	549	30 14 01	23.431
510	26 01 00	22.583	550	30 25 00	23.452
511	26 11 21	22.605	551	30 36 01	23.473
512	26 21 44	22.627	552	30 47 04	23.495
513	26 31 69	22.650	553	30 58 09	23.516
514	26 41 96	22.672	554	30 69 16	23.537
515	26 52 25	22.694	555	30 80 25	23.558
516	26 62 56	22.716	556	30 91 36	23.580
577	26 72 89	22.738	557	31 02 49	23.601
518	26 83 24	22.760	558	31 13 64	23.622
519	26 93 61	22.782	559	31 24 81	23.643
520	27 04 00	22.804	560	31 36 00	23.664

Table I Squares and square roots

Number	Square	Square Root	Number	Square	Square Root
561	31 47 21	23.685	601	36 12 01	24.515
562	31 58 44	23.707	602	36 24 04	24.536
563	31 69 69	23.728	603	36 36 09	24.556
564	31 80 96	23.749	604	36 48 16	24.576
565	31 92 25	23.770	605	36 60 25	24.597
566	32 03 56	23.791	606	36 72 36	24.617
567	32 14 89	23.812	607	36 84 49	24.637
568	32 26 24	23.833	608	36 96 64	24.658
569	32 37 61	23.854	609	37 08 81	24.678
570	32 49 00	23.875	610	37 21 00	24.698
571	32 60 41	23.896	611	37 33 21	24.718
572	32 71 84	23.917	612	37 45 44	24.739
573	32 83 29	23.937	613	37 57 69	24.759
574	32 94 76	23.958	614	37 69 96	24.779
575	33 06 25	23.979	615	37 82 25	24.799
576	33 17 76	24.000	616	37 94 56	24.819
577	33 29 29	24.021	617	38 06 89	24.839
578	33 40 84	24.042	618	38 19 24	24.860
579	33 52 41	24.062	619	38 31 61	24.880
580	33 64 00	24.083	620	38 44 00	24.900
581	33 75 61	24.104	621	38 56 41	24.920
582	33 87 24	24.125	622	38 68 84	24.940
583	33 98 89	24.145	623	38 81 29	24.960
584	34 10 56	24.166	624	38 93 76	24.980
585	34 22 25	24.187	625	39 06 25	25.000
586	34 33 96	24.207	626	39 18 76	25.020
587	34 45 69	24.228	627	39 31 29	25.040
588	34 57 44	24.249	628	39 43 84	25.060
589	34 69 21	24.269	629	39 56 41	25.080
590	34 81 00	24.290	630	39 69 00	25.100
591	34 92 81	24.310	631	39 81 61	25.120
592	35 04 64	24.331	632	39 94 24	25.140
593	35 16 49	24.352	633	40 06 89	25.159
594	35 28 36	24.372	634	40 19 56	25.179
595	35 40 25	24.393	635	40 32 25	25.199
596	35 52 16	24.413	636	40 44 96	25.219
597	35 64 09	24.434	637	40 57 69	25.239
598	35 76 04	24.454	638	40 70 44	25.259
599	35 88 01	24.474	639	40 83 21	25.278
600	36 00 00	24.495	640	40 96 00	25.298

Table I   Squares and square roots

Number	Square	Square Root	Number	Square	Square Root
641	41 08 81	25.318	681	46 37 61	26.096
642	41 21 64	25.338	682	46 51 24	26.115
643	41 34 49	25.357	683	46 64 89	26.134
644	41 47 36	25.377	684	46 78 56	26.153
645	41 60 25	25.397	685	46 92 25	26.173
646	41 73 16	25.417	686	47 05 96	26.192
647	41 86 09	25.436	687	47 19 69	26.211
648	41 99 04	25.456	688	47 33 44	26.230
649	42 12 01	25.475	689	47 47 21	26.249
650	42 25 00	25.495	690	47 61 00	26.268
651	42 38 01	25.515	691	47 74 81	26.287
652	42 51 04	25.534	692	47 88 64	26.306
653	42 64 09	25.554	693	48 02 49	26.325
654	42 77 16	25.573	694	48 16 36	26.344
655	42 90 25	25.593	695	48 30 25	26.363
656	43 03 36	25.612	696	48 44 16	26.382
657	43 16 49	25.632	697	48 58 09	26.401
658	43 29 64	25.652	698	48 72 04	26.420
659	43 42 81	25.671	699	48 86 01	26.439
660	43 56 00	25.690	700	49 00 00	26.458
661	43 69 21	25.710	701	49 14 01	26.476
662	43 82 44	25.729	702	49 28 04	26.495
663	43 95 69	25.749	703	49 42 09	26.514
664	44 08 96	25.768	704	49 56 16	26.533
665	44 22 25	25.788	705	49 70 25	26.552
666	44 35 56	25.807	706	49 84 36	26.571
667	44 48 89	25.826	707	49 98 49	26.589
668	44 62 24	25.846	708	50 12 64	26.608
669	44 75 61	25.865	709	50 26 81	26.627
670	44 89 00	25.884	710	50 41 00	26.646
671	45 02 41	25.904	711	50 55 21	26.665
672	45 15 84	25.923	712	50 69 44	26.683
673	45 29 29	25.942	713	50 83 69	26.702
674	45 42 76	25.962	714	50 97 96	26.721
675	45 56 25	25.981	715	51 12 25	26.739
676	45 69 76	26.000	716	51 26 56	26.758
677	45 83 29	26.019	717	51 40 89	26.777
678	45 96 84	26.038	718	51 55 24	26.796
679	46 10 41	26.058	719	51 69 61	26.814
680	46 24 00	26.077	720	51 84 00	26.833

Table I   Squares and square roots

Number	Square	Square Root	Number	Square	Square Root
721	51 98 41	26.851	761	57 91 21	27.586
722	52 12 84	26.870	762	58 06 44	27.604
723	52 27 29	26.889	763	58 21 69	27.622
724	52 41 76	26.907	764	58 36 96	27.641
725	52 56 25	26.926	765	58 52 25	27.659
726	52 70 76	26.944	766	58 67 56	27.677
727	52 85 29	26.963	767	58 82 89	27.695
728	52 99 84	26.981	768	58 98 24	27.713
729	53 14 41	27.000	769	59 13 61	27.731
730	53 29 00	27.019	770	59 29 00	27.749
731	53 43 61	27.037	771	59 44 41	27.767
732	53 58 24	27.055	772	59 59 84	27.785
733	53 72 89	27.074	773	59 75 29	27.803
734	53 87 56	27.092	774	59 90 76	27.821
735	54 02 25	27.111	775	60 06 25	27.839
736	54 16 96	27.129	776	60 21 76	27.857
737	54 31 69	27.148	777	60 37 29	27.875
738	54 46 44	27.166	778	60 52 84	27.893
739	54 61 21	27.185	779	60 68 41	27.911
740	54 76 00	27.203	780	60 84 00	27.928
741	54 90 81	27.221	781	60 99 61	27.946
742	55 05 64	27.240	782	61 15 24	27.964
743	55 20 49	27.258	783	61 30 89	27.982
744	55 35 36	27.276	784	61 46 56	28.000
745	55 50 25	27.295	785	61 62 25	28.018
746	55 65 16	27.313	786	61 77 96	28.036
747	55 80 09	27.331	787	61 93 69	28.054
748	55 95 04	27.350	788	62 09 44	28.071
749	56 10 01	27.368	789	62 25 21	28.089
750	56 25 00	27.386	790	62 41 00	28.107
751	56 40 01	27.404	791	62 56 81	28.125
752	56 55 04	27.423	792	62 72 64	28.142
753	56 70 09	27.441	793	62 88 49	28.160
754	56 85 16	27.459	794	63 04 36	28.178
755	57 00 25	27.477	795	63 20 25	28.196
756	57 15 36	27.495	796	63 36 16	28.213
757	57 30 49	27.514	797	63 52 09	28.231
758	57 45 64	27.532	798	63 68 04	28.249
759	57 60 81	27.550	799	63 84 01	28.267
760	57 76 00	27.568	800	64 00 00	28.284

Table I   Squares and square roots

Number	Square	Square Root	Number	Square	Square Root
801	64 16 01	28.302	841	70 72 81	29.000
802	64 32 04	28.320	842	70 89 64	29.017
803	64 48 09	28.337	843	71 06 49	29.034
804	64 64 16	28.355	844	71 23 36	29.052
805	64 80 25	28.373	845	71 40 25	29.069
806	64 96 36	28.390	846	71 57 16	29.086
807	65 12 49	28.408	847	71 74 09	29.103
808	65 28 64	28.425	848	71 91 04	29.120
809	65 44 81	28.443	849	72 08 01	29.138
810	65 61 00	28.460	850	72 25 00	29.155
811	65 77 21	28.478	851	72 42 01	29.172
812	65 93 44	28.496	852	72 59 04	29.189
813	66 09 69	28.513	853	72 76 09	29.206
814	66 25 96	28.531	854	72 93 16	29.223
815	66 42 25	28.548	855	73 10 25	29.240
816	66 58 56	28.566	856	73 27 36	29.257
817	66 74 89	28.583	857	73 44 49	29.275
818	66 91 24	28.601	858	73 61 64	29.292
819	67 07 61	28.618	859	73 78 81	29.309
820	67 24 00	28.636	860	73 96 00	29.326
821	67 40 41	28.653	861	74 13 21	29.343
822	67 56 84	28.671	862	74 30 44	29.360
823	67 73 29	28.688	863	74 47 69	29.377
824	67 89 76	28.705	864	74 64 96	29.394
825	68 06 25	28.723	865	74 82 25	29.411
826	68 22 76	28.740	866	74 99 56	29.428
827	68 39 29	28.758	867	75 16 89	29.445
828	68 55 84	28.775	868	75 34 24	29.462
829	68 72 41	28.792	869	75 51 61	29.479
830	68 89 00	28.810	870	75 69 00	29.496
831	69 05 61	28.827	871	75 86 41	29.513
832	69 22 24	28.844	872	76 03 84	29.530
833	69 38 89	28.862	873	76 21 29	29.547
834	69 55 56	28.879	874	76 38 76	29.563
835	69 72 25	28.896	875	76 56 25	29.580
836	69 88 96	28.914	876	76 73 76	29.597
837	70 05 69	28.931	877	76 91 29	29.614
838	70 22 44	28.948	878	77 08 84	29.631
839	70 39 21	28.965	879	77 26 41	29.648
840	70 56 00	28.983	880	77 44 00	29.665

Table I   Squares and square roots

Number	Square	Square Root	Number	Square	Square Root
881	77 61 61	29.682	921	84 82 41	30.348
882	77 79 24	29.698	922	85 00 84	30.364
883	77 96 89	29.715	923	85 19 29	30.381
884	78 14 56	29.732	924	85 37 76	30.397
885	78 32 25	29.749	925	85 56 25	30.414
886	78 49 96	29.766	926	85 74 76	30.430
887	78 67 69	29.783	927	85 93 29	30.447
888	78 85 44	29.799	928	86 11 84	30.463
889	79 03 21	29.816	929	86 30 41	30.480
890	79 21 00	29.833	930	86 49 00	30.496
891	79 38 81	29.850	931	86 67 61	30.512
892	79 56 64	29.866	932	86 86 24	30.529
893	79 74 49	29.883	933	87 04 89	30.545
894	79 92 36	29.900	934	87 23 56	30.561
895	80 10 25	29.916	935	87 42 25	30.578
896	80 28 16	29.933	936	87 60 96	30.594
897	80 46 09	29.950	937	87 79 69	30.610
898	80 64 04	29.967	938	87 98 44	30.627
899	80 82 01	29.983	939	88 17 21	30.643
900	81 00 00	30.000	940	88 36 00	30.659
901	81 80 01	30.017	941	88 54 81	30.676
902	81 36 04	30.033	942	88 73 64	30.692
903	81 54 09	30.050	943	88 92 49	30.708
904	81 72 16	30.067	944	89 11 36	30.725
905	81 90 25	30.083	945	89 30 25	30.741
906	82 08 36	30.100	946	89 49 16	30.757
907	82 26 49	30.116	947	89 68 09	30.773
908	82 44 64	30.133	948	89 87 04	30.790
909	82 62 81	30.150	949	90 06 01	30.806
910	82 81 00	30.166	950	90 25 00	30.822
911	82 99 21	30.183	951	90 44 01	30.838
912	83 17 44	30.199	952	90 63 04	30.854
913	83 35 69	30.216	953	90 82 09	30.871
914	83 53 96	30.232	954	91 01 16	30.887
915	83 72 25	30.249	955	91 20 25	30.903
916	83 90 56	30.265	956	91 39 36	30.919
917	84 08 89	30.282	957	91 58 49	30.935
918	84 27 24	30.299	958	91 77 64	30.952
919	84 45 61	30.315	959	91 96 81	30.968
920	84 64 00	30.332	960	92 16 00	30.984

Table I Squares and square roots

Number	Square	Square Root	Number	Square	Square Root
961	92 35 21	31.000	981	96 23 61	31.321
962	92 54 44	31.016	982	96 43 24	31.337
963	92 73 69	31.032	983	96 62 89	31.353
964	92 92 96	31.048	984	96 82 56	31.369
965	93 12 25	31.064	985	97 02 25	31.385
966	93 31 56	31.081	986	97 21 96	31.401
967	93 50 89	31.097	987	97 41 69	31.417
968	93 70 24	31.113	988	97 61 44	31.432
969	93 89 61	31.129	989	97 81 21	31.448
970	94 09 00	31.145	990	98 01 00	31.464
971	94 28 41	31.161	991	98 20 81	31.480
972	94 47 84	31.177	992	98 40 64	31.496
973	94 67 29	31.193	993	98 60 49	31.512
974	94 86 76	31.209	994	98 80 36	31.528
975	95 06 25	31.225	995	99 00 25	31.544
976	95 25 76	31.241	996	99 20 16	31.559
977	95 45 29	31.257	997	99 40 09	31.575
978	95 64 84	31.273	998	99 60 04	31.591
979	95 84 41	31.289	999	99 80 01	31.607
980	96 04 00	31.305	1000	100 00 00	31.623

Table II   Random digits

03991	10461	93716	16894	98953	73231	39528	72484	82474	25593
38555	95554	32886	59780	09958	18065	81616	18711	53342	44276
17546	73704	92052	46215	15917	06253	07586	16120	82641	22820
32643	52861	95819	06831	19640	99413	90767	04235	13574	17200
69572	68777	39510	35905	85244	35159	40188	28193	29593	88627
24122	66591	27699	06494	03152	19121	34414	82157	86887	55087
61196	30231	92962	61773	22109	78508	63439	75363	44989	16822
30532	21704	10274	12202	94205	20380	67049	09070	93399	45547
03788	97599	75867	20717	82037	10268	79495	04146	52162	90286
48228	63379	85783	47619	87481	37220	91704	30552	04737	21031
88618	19161	41290	67312	71857	15957	48545	35247	18619	13674
71299	23853	05870	01119	92784	26340	75122	11724	74627	73707
27954	58909	82444	99005	04921	73701	92904	13141	32392	19763
80863	00514	20247	81759	45197	25332	69902	63742	78464	22501
33564	60780	48460	85558	15191	18782	94972	11598	62095	36787
90899	75754	60833	25983	01291	41349	19152	00023	12302	80783
78038	70267	43529	06318	38384	74761	36024	00867	76378	41605
55986	66485	88722	56736	66164	49431	94458	74284	05041	49807
87539	08823	94813	31900	54155	83436	54158	34243	46978	35482
16818	60311	74457	90561	72848	11834	75051	93029	47665	64382
34677	58300	74910	64345	19325	81549	60365	94653	35075	33949
45305	07521	61318	31855	14413	70951	83799	42402	56623	34442
59747	67277	76503	34513	39663	77544	32960	07405	36409	83232
16520	69676	11654	99893	02181	68161	19322	53845	57620	52606
68652	27376	92852	55866	88448	03584	11220	94747	07399	37408
79375	95220	01159	63267	10622	48391	31751	57260	68980	05339
33521	26665	55823	47641	86225	31704	88492	99382	14454	04504
59589	49067	66821	41575	49767	04037	30934	47744	07481	83828
20554	91409	96277	48257	50816	97616	22888	48893	27499	98748
59404	72059	43947	51680	43852	59693	78212	16993	35902	91386
42614	29297	01918	28316	25163	01889	70014	15021	68971	11403
34994	41374	70071	14736	65251	07629	37239	33295	18477	65622
99385	41600	11133	07586	36815	43625	18637	37509	14707	93997
66497	68646	78138	66559	64397	11692	05327	82162	83745	22567
48509	23929	27482	45476	04515	25624	95096	67946	16930	33361
15470	48355	88651	22596	83761	60873	43253	84145	20368	07126
20094	98977	74843	93413	14387	06345	80854	09279	41196	37480
73788	06533	28597	20405	51321	92246	80088	77074	66919	31678
60530	45128	74022	84617	72472	00008	80890	18002	35352	54131
44372	15486	65741	14014	05466	55306	93128	18464	79982	68416
18611	19241	66083	24653	84609	58232	41849	84547	46850	52326
58319	15997	08355	60860	29735	47762	46352	33049	69248	93460
61199	67940	55121	29281	59076	07936	11087	96294	14013	31792
18627	90872	00911	98936	76355	93779	52701	08337	56303	87315
00441	58997	14060	40619	29549	69616	57275	36898	81304	48585
32624	68691	14845	46672	61958	77100	20857	73156	70284	24326
65961	73488	41839	55382	17267	70943	15633	84924	90415	93614
20288	34060	39685	23309	10061	68829	92694	48297	39904	02115
59362	95938	74416	53166	35208	33374	77613	19019	88152	00080
99782	93478	53152	67433	35663	52972	38688	32486	45134	63545

Reprinted by permission from INTRODUCTION TO MATHEMATICAL STATISTICS, 3rd edition, by Paul G. Hoel, © 1962 by John Wiley & Sons, Inc.

Table II   Random digits

27767	43584	85301	88977	29490	69714	94015	64874	32444	48277
13025	14338	54066	15243	47724	66733	74108	88222	88570	74015
80217	36292	98525	24335	24432	24896	62880	87873	95160	59221
10875	62004	90391	61105	57411	06368	11748	12102	80580	41867
54127	57326	26629	19087	24472	88779	17944	05600	60478	03343
60311	42824	37301	42678	45990	43242	66067	42792	95043	52680
49739	71484	92003	98086	76668	73209	54244	91030	45547	70818
78626	51594	16453	94614	39014	97066	30945	57589	31732	57260
66692	13986	99837	00582	81232	44987	69170	37403	86995	90307
44071	28091	07362	97703	76447	42537	08345	88975	35841	85771
59820	96163	78851	16499	87064	13075	73035	41207	74699	09310
25704	91035	26313	77463	55387	72681	47431	43905	31048	56699
22304	90314	78438	66276	18396	73538	43277	58874	11466	16082
17710	59621	15292	76139	59526	52113	53856	30743	08670	84741
25852	58905	55018	56374	35824	71708	30540	27886	61732	75454
46780	56487	75211	10271	36633	68424	17374	52003	70707	70214
59849	96169	87195	46092	26787	60939	59202	11973	02902	33250
47670	07654	30342	40277	11049	72049	83012	09832	25571	77628
94304	71803	73465	09819	58869	35220	09504	96412	90193	79568
08105	59987	21437	36786	49226	77837	98524	97831	65704	09514
64281	61826	18555	64937	64654	25843	41145	42820	14924	39650
66847	70495	32350	02985	01755	14750	48968	38603	70312	05682
72461	33230	21529	53424	72877	17334	39283	04149	90850	64618
21032	91050	13058	16218	06554	07850	73950	79552	24781	89683
95362	67011	06651	16136	57216	39618	49856	99326	40902	05069
49712	97380	10404	55452	09971	59481	37006	22186	72682	07385
58275	61764	97586	54716	61459	21647	87417	17198	21443	41808
89514	11788	68224	23417	46376	25366	94746	49580	01176	28838
15472	50669	48139	36732	26825	05511	12459	91314	80582	71944
12120	86124	51247	44302	87112	21476	14713	71181	13177	55292
95294	00556	70481	06905	21785	41101	49386	54480	23604	23554
66986	34099	74474	20740	47458	64809	06312	88940	15995	69321
80620	51790	11436	38072	40405	68032	60942	00307	11897	92674
55411	85667	77535	99892	71209	92061	92329	98932	78284	46347
95083	06783	28102	57816	85561	29671	77936	63574	31384	51924
90726	57166	98884	08583	95889	57067	38101	77756	11657	13897
68984	83620	89747	98882	92613	89719	39641	69457	91339	22502
36421	16489	18059	51061	67667	60631	84054	40455	99396	63680
92638	40333	67054	16067	24700	71594	47468	03577	57649	63266
21036	82808	77501	97427	76479	68562	43321	31370	28977	23896
13173	33365	41468	85149	49554	17994	91178	10174	29420	90438
86716	38746	94559	37559	49678	53119	98189	81851	29651	84215
92581	02262	41615	70360	64114	58660	96717	54244	10701	41393
12470	56500	50273	93113	41794	86861	39448	93136	25722	08564
01016	00857	41396	80504	90670	08289	58137	17820	22751	36518
34030	60726	25807	24260	71529	78920	47648	13885	70669	93406
50259	46345	06170	97965	88302	98041	11947	56203	19324	20504
73959	76145	60808	54444	74412	81105	69181	96845	38525	11600
46874	37088	80940	44893	10408	3622.	14004	23153	69249	05747
60883	52109	19516	90120	46759	71643	62342	07589	08899	05985